G. R Burton

ANATOMY AND PHYSIOLOGY.

A TREATISE

ON

ANATOMY AND PHYSIOLOGY,

BY

Wm P. ESREY, M. D.

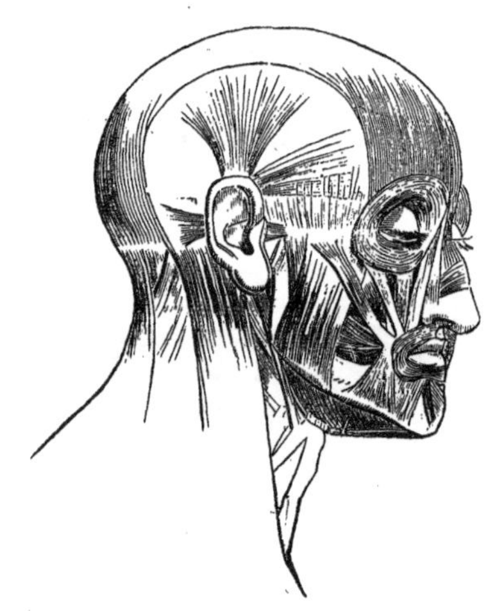

WITH THIRTY ILLUSTRATIONS.

PHILADELPHIA:
PUBLISHED BY RADEMACHER & SHEEK, No. 239 ARCH STREET.
NEW YORK: WM. RADDE, 322 BROADWAY.
ST. LOUIS: J. G. WESSELHŒFT, 64 NORTH FOURTH ST.
BOSTON: OTIS CLAPP, 23 SCHOOL ST.
1851.

PREFACE.

The following Treatise was prepared to accompany a work on Domestic Medicine, and without any design of publishing it in a separate form. The Publishers, however, have thought proper to issue it separately, believing that the multiplication of books of the kind, especially when presented in a cheap and condensed form, cannot be other than productive of good results, as a greater number of persons will thereby be induced to read and study them. Some knowledge of the structure and functions of our organization is almost indispensable, in order to the proper preservation and enjoyment of health, without which life itself frequently becomes burdensome, and with which contentment may be found in the humblest cottage.

To insure accuracy, the latest and best works on the subjects have been carefully consulted, and the views of the most distinguished Anatomists and Physiologists given, as far as the limits allowed would admit of it.

Anatomy and Physiology.

THE growing interest manifested in all classes to become acquainted with the structure and formation of the human organism is sufficient excuse, if any be needed, for the introduction of some general remarks on anatomy and physiology into a work of this kind. It is indeed important that those laymen who occasionally prescribe for disease should have some acquaintance with these subjects, to enable them to discriminate with greater accuracy between different morbid conditions, and to distinguish really dangerous symptoms from those of but little consequence.

No pretension is made to originality of matter in the preparation of this treatise, but the best standard text-books on the subjects have been freely consulted; and the chief aim has been to condense into a small compass as much useful matter as possible, though from the very limited space to be occupied, and the wide field to go over, the information on many points must necessarily be brief and imperfect.

The subjects being very intimately allied—*Anatomy* being the science which investigates the structure and organization of living beings, *Physiology* the science which treats of the actions or functions peculiar to living organized beings while in health—it has been thought best to consider them together, rather than under separate heads, as is done in all the more systematic treatises.

Comparative Anatomy treats of the *structure* of the lower orders of animals.

Comparative Physiology considers the *functions* of the inferior animals.

Vegetable Physiology treats exclusively of plants. It is to *Human Anatomy* and *Physiology* that this treatise will be principally devoted.

The substances in nature are divisible into two chief classes:—the *organized*, or those which possess separate parts or organs suited for the performance of certain actions or functions, and the *inorganic*, or such as are without this arrangement. The former includes animals and vegetables, the latter the earths, metals, &c. All organized bodies have their origin from parents; they must spring from beings like themselves, and they are controlled by *vitality*, or the *principle of life*. The several separate organs have a mutual dependence on each other, and to preserve the harmony of the whole, the integrity of each part must be preserved.

Organized bodies have likewise a definite *shape* and *size*, and an allotted time to live; and after death they are decomposed, and pass into the simple combination of the inorganic elements.

Inorganic bodies, on the contrary, do not spring from parents; they are not born; but they have their origin in, and are governed by certain fixed and invariable natural laws. The particles of which they consist are in a state of aggregation only. And their growth, or increase of size, or change of shape, takes place by the accretion of matter to their surfaces. The parts of which an inorganic body is composed are all alike in structure and properties, and may exist as well in detached portions, or when broken in pieces, as in large masses. And each body consists wholly of matter, either in the solid, liquid, or gaseous form.

Organized bodies, on the other hand, are always composed of a combination of solids and fluids; they differ in character and properties, and are arranged into organs, so as to form of the whole a single system.

They also increase in size by a process called nutrition, which consists in imbibing substances from without, and changing them, by means of internal organs, to their own nature.

In respect to the *chemical* character of inorganic and organic bodies, great difference exists. All the substances considered elementary, amounting to about sixty, are found in the mineral kingdom. In the organized, only a few of these elements are met with—about seventeen—and of these but four are considered essential, viz. carbon, hydrogen, oxygen, and nitrogen; two of which, at least, will be found in every organic compound. The other simple elements exist in smaller quantities, and may be considered non-essential: they are, sulphur, phosphorus, chlorine, sodium, calcium, potassium, magnesium, silicon, aluminum, iron, manganese, iodine, and bromium. The last two are found principally in marine plants and animals. The composition of inorganic bodies is more simple, some consisting of but one element; and when composed of more the combination rarely exceeds three. Organized bodies, on the contrary, in most instances consist of three or four elements; the simplest vegetable consists of a union of oxygen, carbon, and hydrogen; and the simplest animal of oxygen, hydrogen, carbon, and nitrogen.

As both animal and vegetables are included under the head of organized bodies, it is necessary to point out some of the distinctive characters between these kingdoms. Nutrition, or the capacity of assimilating foreign matters to their own substance for purposes of development and growth, and reproduction, or the power of producing a living being similar to themselves, are properties common to each. Besides these there are other properties peculiar to animals, and hence called *animal* properties or functions. These are *sensation* and *voluntary motion*.

Though plants are not destitute of motion—their roots seeking the most nutritious soil; the leaves and branches turning spontaneously to the light; the internal circulation of the sap; and some, as the sensitive plant, even seem to perform movements almost indicative of sensibility—yet these are quite different from the motions of animals, not being the result of consciousness. They result from physical changes produced directly in the part of the plant touched, and strictly organic; and are indications of *irritability*, but not of *sensibility*. A distinction is to be made between these qualities. In the lower orders of animals, indeed, the distinction is more difficult; naturalists having been for a long time undecided whether the zoophites—sponge, coral, &c.—should be considered as animals or vegetables. They are as firmly attached to the soil as the latter, and like them receive their nourishment from without. Some species of plants, on the other hand, as the *Fucus natans*, or gulf weed, lives and fructifies on the water, and are constantly borne about by the waves. This motion, however, is very different from that of animals; it is entirely passive.

Animals likewise differ from vegetables very materially with respect to *nutrition.* In the former a stomach is necessary, in order to receive the food, which is generally crude and unfitted for absorption, and prepare it for the nutrition of the individual by a process termed digestion. The nourishment of plants is derived from inorganic substances, from the excretions of animals, and from decaying organic matter, while animals can only be nourished by organized substances, either animal or vegetable. The absorption of nutritive matters by the latter is from without, by means of the roots; that of the former from within, by vessels situated in the lining membrane of the alimentary canal.

Differences also exist in the functions of reproduction; in the animal, volition is required in almost every stage of the process; in vegetables, on the contrary, the whole is effected without the exercise of volition or consciousness.

The proportion of fluids to solids is, likewise, much greater in animals than in vegetables, which accounts for decomposition taking place much sooner in the former than in the latter.

All organized bodies are composed chiefly of carbon, oxygen, hydrogen, and nitrogen, with alkaline and earthy salts. Vegetables mostly consist of the three first of these elements, and nitrogen is combined with them in the animal only. Or, when this latter element is found in vegetables, it is in limited quantity, and generally confined to one part. It is this difference of composition which gives to animal substances the peculiar smell when burning by which they are readily distinguished from vegetables.

Elementary Composition of Man.

The human body is composed of fluids and solids; the blood, chyle, lymph, and the various secretions constitute the former; and the various textures, as the bones, muscles, viscera, &c., the latter. Water is one of the most important constituents of the human body; it is in large proportion in all the fluids, and also in the solids, and gives to the latter softness and flexibility.

The proportion of fluids is far greater than that of the solids, being in the proportion of six or nine to one, according to the estimate of different observers, it being exceedingly difficult to arrive at an exact estimate. *Chaussier* found a dead body which weighed one hundred and twenty pounds; after being dried in an oven, it weighed only twelve. And a perfectly dry mummy in the possession of *Blumenbach*, which contained all the viscera, weighed but seven pounds and a half.

There exists in organic structures a class of compounds, called *proximate principles*, or *organic elements*, which consist of two or more of the elementary substances, combined in definite proportions.

The chief of these are: albumen, caseine, fibrine, gelatine, chondrine, elaine, stearine, margarine, hæmatosine, and globuline.

Albumen.—Is found in two forms, *fluid* and *concrete;* the former, which is met with in the white of egg, is colorless and transparent, without smell or taste, and is coagulable by heat, acids, and corrosive sublimate; it is found in the blood, lymph, and chyle. The latter, concrete or solid albumen, is white, tasteless and elastic; it is insoluble in water, alcohol, or oil, but readily soluble in alkalies; it is found in the brain, spinal cord, and nerves, and in the mucous membranes. Hair, nails and horn, also, consist principally of albumen. It is one of the most common of the organic constituents.

Caseine.—Exists abundantly in milk, and is the basis of cheese. It may be obtained by allowing milk to remain at rest till it is coagulated, taking off the cream, then washing the clot in water and drying it. It is readily coagulated by the action of *rennet;* this is owing to the pepsin contained in the latter. Caseine is white, insipid, and inodorous, insoluble in water, but readily soluble in the alkalies, ammonia especially. It contains sulphur. It has many properties analogous with albumen.

Fibrine.—This principle exists in the chyle, lymph, and blood, in solution; it also forms the basis of the muscles, where it is found in the solid form, and is one of the most abundant of the animal substances. It may be obtained by beating blood with a stick as it is flowing from a vein, or by washing a clot repeatedly in clean water so as to dissolve out the coloring matter.

Fibrine is solid, white, flexible, and slightly elastic, insipid, inodorous, and heavier than water; it is insoluble in water, alcohol, and acids, but soluble in caustic potassa. Chemically speaking, it does not differ essentially from albumen; the chief variation is physiologically, in the spontaneous coagulation of the fibrine; in coagulating, the fibres assume a definite arrangement, crossing each other in all directions. It constitutes the buffy coat of the blood. In the reparation of injured parts it is thrown out from the blood vessels as a secretion, and becomes organized. It is often called coagulable lymph under such circumstances.

Gelatine.—Is the chief ingredient of the cellular tissue, skin, tendons, cartilages, and ligaments; it also enters largely into the composition of bones. It may be obtained by boiling any of these substances for some time in water; clarifying the concentrated solution; allowing it to cool, and then drying the substance obtained in the air. When dry and hard, it is called *glue;* in a liquid form *jelly.* Gelatine is soluble in hot water, in acids and alkalies. It is insoluble in cold water, alcohol and ether, and has a strong affinity for tannin. The process of tanning leather results from the combination of tannin with gelatine. The air-bags of fishes consist of pure gelatine, known as isinglass. The article known as portable soup consists of dried gelatine seasoned with spices. Under the form of *glue* and *size,* gelatine is extensively used in the arts, its adhesive properties rendering it valuable.

Chondrine.—Resembles gelatine, but does not unite with tannin, and is precipitated by acetic acid, acetate of lead, alum, and protosulphate of iron. It is obtained by boiling the cartilages, and allowing the solution to cool.

Elaine, stearine, and *margarine,* are the proximate principles of fat. The first is fluid at ordinary temperatures; the second is fluid, and is the chief ingredient of vegetable and animal suet, and of fat and butter; it is but sparingly present in human fat. The third is of medium consistency.

Hæmatosine—Is the red coloring matter of the blood, contained in a capsule which is composed of *globuline.*

There are other organic elements, called *secondary* organic compounds, as *urea,* cholestrine, pepsine, sugar of milk, &c., which are excretions of particular organs. They will be treated of under the head of secretions.

Of the *inorganic, ultimate,* or *chemical elements* it has been already remarked that oxygen, hydrogen, carbon, and nitrogen, are the most essential. Besides these there are phosphorus, lime, sulphur, iron, manganese, silicum, chlorine, sodium, magnesium, &c.

Oxygen—Is distributed throughout nearly all the solids and fluids. It is indispensable to life, and a supply is constantly furnished from the atmosphere. It is mostly found combined with other bodies, often with carbon in the form of carbonic acid.

Hydrogen —This gas is found universally distributed throughout the animal kingdom. It is contained in all the fluids and in most of the solids, and is generally in combination with carbon. It has been found pure in the intestines, as well as combined with carbon and sulphur.

In the form of sulphuretted hydrogen gas, it has an offensive smell. The flatulence emitted from the intestines mostly consists of this combination.

Carbon.—Is met with in both fluids and solids in various forms. Generally it exists under the form of carbonic acid. It is emitted by all animals in the act of expiration. In animal bodies it mostly exists in combination with alkalies, or earths, though it has been found uncombined.

Nitrogen.—Exists extensively in animal substances. Its prevalence is so general that it frequently serves as a test to distinguish them from vegetables. It is this principle that gives to animal substances their peculiar smell when burning.

Phosphorus—Occurs in combination with oxygen—*phosphoric acid*—in many animal substances, both solids and fluids. Combined with earthy matters this acid forms one of the chief ingredients of bones. It is likewise combined in other parts with potassa, soda, ammonia, and magnesia.

Calcium or *lime*, is found in the state of oxide of lime only in animals. It is generally in the form of phosphate or carbonate. This earth constitutes the hard parts of animals, as bone, &c.

Sulphur—Is sparingly met with in animal fluids and solids, and is always combined with oxygen, and united to lime, soda, or potash. In the lower part of the intestines, and also as an exhalation from fetid ulcers, it is met with in the form of sulphuretted hydrogen gas.

Iron—Is found in the blood; in bile and in milk. It is the coloring matter of the red globules of blood.

Manganese.—This substance has been found along with iron in the ashes of the hair in a state of oxide.

Silicum.—Exists in the hair and urine.

Chlorine.—Is contained in most of the animal fluids; it is generally combined with hydrogen, forming muriatic acid which is united with soda. *Muriatic acid* in a free state also exists in the stomach.

Sodium.—As an oxide soda is found in all of the fluids. It is likewise united to albumen. Most frequently, however, it is combined with the phosphoric and muriatic acids; sometimes also with the lactic, sulphuric, and carbonic acid.

Potassium—Is likewise found in animal fluids united with acids. It is, however, more abundant in vegetables.

Magnesium—Exists in the form of an oxide, *magnesia*, sparingly in the bones, and in some other parts. It is always combined with phosphoric acid.

Out of the *proximate principles* described, the various tissues of the human body are formed. The different solid parts are arranged in a variety of ways: and of these the principal are in *filaments*, or elementary fibres, *tissues*, *organs*, *apparatusses*, and *systems*.

By a *filament* is understood the elementary solid, consisting of minute particles of matter arranged in a row.

A *fibre* consists of a number of filaments united together and enclosed in a sheath.

By the term *tissue* is generally meant a particular arrangement of fibres.

Organs are formed by the union of tissues—the liver, stomach, &c., are examples.

An *apparatus* consists of the union of several different organs to accomplish one end—as the biliary apparatus, consisting of the liver, gall bladder, &c., all of which aid in the secretion of food.

A *system* is composed of a number of similar organs, united for one end—as the muscular or nervous system.

Anatomists define a solid to be, a body the particles of which adhere to each other, so that they will not separate by their own weight, but require the application of some external force to effect disunion. And they likewise divide the various solids of which the human body is composed, into the following varieties, viz. bone, cartilage, muscle, ligament, vessel, nerve, ganglion, follicle, gland, membrane, cellular membrane, and viscus.

Bone.—This is the hardest of the solids, and forms the skeleton, serving as a protection to various important organs and for attachments to muscles.

Cartilage.—Ranks next to bone in hardness; it is white and elastic, covers the articulating extremities of bone to facilitate their movements; it serves also in some cases, as in the ribs, to prolong the bones; in the fœtus it is a substitute for bone There are several varieties of cartilage.

Muscles.—What is called flesh in animals consists of muscles. They are the agents of all movements, and are composed of bundles of red and contractile fibres, extending from one bone to another.

Ligaments.—Are cords or bands, exceedingly tough and difficult to tear. They serve to connect different parts to each other, as the bones and muscles. By some they are divided into two varieties—those which connect the joints, and those attached to the muscles,—the tendons and aponeuroses.

Vessels.—These are in the form of canals or tubes, and serve to carry on the circulation of the various fluids. And are called sanguineous, chyliferous, lymphatic, &c., according to the nature of the fluid they carry.

Nerves.—The nerves are solid cords consisting of numerous fasciculi or bundles. They are connected with the brain, spinal marrow, or great sympathetic, and their function is to convey impressions to the nervous centres; and to endow each part with vitality. There are two chief divisions of nerves—those of the brain and spinal marrow, and the organic or great sympathetic.

Ganglions.—A ganglion is a knot situated in the course of a nerve and formed apparently by an interlacing of filaments. The term is likewise applied to a similar interlacing of lymphatic vessels.

Follicles or crypts.—Are small membranous vesicles seated in the substance of the skin or mucous membrane.

Their office is to secrete a fluid to lubricate these parts. It is the secretion from these follicles chiefly that keeps the skin soft, and gives to it its oily character.

Glands.—Are likewise secretory organs, but differ from the follicle—their organization being more complicated. The liver, for instance, is a gland, and its office the secretion of bile. The glands of the human body are numerous and diversified in their character.

Membrane.—The membrane is formed by the cellular tissue, and is one of the most important and extensive substances of the body. It is spread out like a web, and serves to form, support, and envelope all the organs, to line the cavities and reservoirs. It is divided into simple and compound. The simple being again subdivided into three varieties, viz. the *serous, mucous,* and *fibrous.* The first of these, the serous, form all the sacs or closed cavities of the body—as those of the chest and abdomen. The compound membranes are formed by the union of the simple, and are divided into *fibro-serous*, *sero-mucous*, and *fibro-mucous*. The pericardium is an example of the first of these; the gall-bladder of the second; and the ureter of the third.

Cellular or *laminated tissue.*—This is a kind of spongy or areolar structure, which encloses all the solids, fills up the spaces between them, serving at the same time as a bond of union and a medium of separation. It will be more fully described presently.

Viscus.—The name viscus is given to those solids which are the most complicated, both as regards texture and use. The brain, lungs, liver, &c., are examples.

The *tissues* have been variously classified: the classification of *Haller*, which reduces them to three primary ones, has been very generally adopted by anatomists and physiologists. These are the *cellular* or *areolar*, the *muscular*, and the *nervous*, out of which all the organs are formed, either from the first alone, or by the union of the last two.

1. *The cellular* or *areolar tissue.*—This tissue is the most simple and abundant of the solids; it exists in all organized beings, and is an element in every solid, with the exception perhaps of the enamel of the teeth, where it has not as yet been detected. It is formed by the interlacing and crossing of numerous fibres or bands, of a delicate whitish color, so as to leave numerous interstices, or areola, which communicate with each other. A proof of this is furnished in anasarca or general dropsy, where the excess of fluid passes readily from one part to another; and by pressing with the finger, the fluid is forced into adjoining parts of the membrane, and a pit is formed which gradually disappears again on removing the pressure. They may also be filled with air, as occurs occasionally in the disease called emphysema; and it is a knowledge of this fact that enables butchers to inflate or *blow* their meat, to give to it a fat appearance. This membrane possesses elasticity and extensibility, but not much vitality; it is composed chiefly of gelatine. During life the interstices are filled with a dilute serous or watery fluid,

which passes from the blood-vessels. It is the excess of this fluid caused by disease that constitutes anasarca or general dropsy.

2. *Muscular tissue*—Consists of an arrangement of exceedingly minute fibres or filaments of a peculiar substance. These fibres are arranged in parallel layers in all the voluntary muscles and a few others; in the involuntary, including those of the alimentary canal, bladder, uterus, &c., they interlace. They are soft, of a grayish or reddish color, and possessed of contractibility or irritability, that is, they move responsive to chemical and mechanical irritants. They are composed principally of fibrine.

Nervous tissues.—This tissue or fibre is of a pulpy consistence; it is composed essentially of albumen united to a fatty matter, and is the organ of sensibility, or for receiving and conveying impressions to the mind. The brain, nerves, &c., are composed of it, and it is not near so generally distributed as the preceding. The ultimate nervous fibre or filament is said to be ten or twelve times larger than that of muscle.

These three varieties of tissues or fibres, by uniting in different proportion, from the first order of solids; and these again by union constitute compound tissues, out of which the various organs, glands, bones, &c., are formed. Thus, for instance, a bone is composed of several tissues, the body being *osseous*, the interior *nervous*, the extremities *cartilaginous*, and the exterior *fibrous*.

The primary form taken by organic matter, as it passes from the state of a proximate principle to that of an organized structure, appears to be a cell; this cell contains another within it, called *nucleus*, which again contains a granular body, called *nucleolus*. Almost all the organic tissues, however unlike, are in the embryo composed of cells which are afterwards developed by a vital process into the various structures that make up the perfect being. Cells in great numbers are found floating in the blood, chyle and lymph, and their development goes on during the life of the organism.

Physical and vital properties of the tissues.—The tissues present striking differences, as well in their properties, as in their anatomical arrangement. These properties are divided into *physical* and *vital.* The physical belong as well to the dead as to the living tissue, and are dependent altogether on the particular arrangement or mode of cohesion of their constituent particles and their chemical composition. The vital properties are those which belong exclusively to the *living* organism, and which terminate with organic life.

The most marked of the physical properties are: *elasticity, flexibility, extensibility,* and *porosity.*

Elasticity is that power by which a tissue reacts after the withdrawal of an extending or compressing force. It is manifested in the yellow ligament, in the middle coat of the arteries, and in the cartilages of the ribs, and the articular faces of the bones.

Extensibility. — Tissues may be extensible without being elastic; those which are so yield only to long continued pressure; an example of this is furnished in the resistance of fibrous membranes to the growth of a tumour.

Flexibility, or the capability of being bent, is witnessed in the white fibrous tissues, which are flexible, and not elastic nor extensible. The tendons are examples.

Porosity.—The tissues even after death are porous, that is, they are permeable by watery fluids. This property is also termed imbibition. The softness of tissues is owing to the watery fluids which fill their pores.

The vital properties of tissues.—These properties which belong only to organic life, manifest themselves by a change of their molecules or elementary constituents, from the application of a stimulus. This change may take place directly, with a visible alteration of the tissue impressed, or indirectly through the intervention of some other organ or tissue with which the impressed tissue may be in connection.

There are two tissues which have these properties, viz. the muscular and the nervous. In nerves it is manifested in three ways: 1) by causing contraction in the muscle supplied by it; 2) by inducing contraction in muscles not supplied by it, through a change in the nervous centres with which it is connected; 3) by exciting sensation.

When seated in *muscle* it is called contractility, and is characteristic of that tissue. Hence there are *two* vital properties,—*sensibility* and *contractility*—the first, seated in the nervous tissue, the second in the muscular. Vegetables having no nervous system cannot be sensible, as this property requires a brain, a particular part of the nervous system, for its exercise. The only *vital* property, therefore, which belongs both to animals and vegetables is better expressed by the term *excitability* or irritability. Plants are not *sensible;* being without consciousness, they are *irritable.* Muscles, when removed from the body, are likewise *irritable*, and not sensible. Physiologists have variously termed this one *vital property*, irritability, excitability, contractility, or incitability.

Of the Bones.

Anatomists generally reckon two hundred and eleven bones in the human skeleton, though the number varies, they being more numerous in youth than in old age; some bones which in the young consist of several pieces, as life advances become united by the ossification of the intervening cartilages. Bones are of a dull white color, hard, and inflexible; by boiling and proper preparation they may be made of an ivory whiteness; when held together by their natural connection of ligaments and cartilages they form a *natural* skeleton; when joined by wires, or other means, an *artificial* skeleton.

The regional division of the skeleton is into the head, trunk, the superior, and the inferior extremities. Twenty-two bones belong to the head, fifty-six to the trunk, sixty-nine to the superior, and sixty-four to the inferior extremities.

The bones are classified as long, thick and flat bones, and are either symmetrical, that is, they consist of two lateral portions exactly alike; or else are in pairs, which have a perfect correspondence with each other.

The symmetrical bones are, the frontal, occipital, sphenoidal, ethmoidal, vomer, inferior maxillary, hyoid, the spinal, and the sternal; all of these are placed in the middle verticle line of the body. The pairs are situated on both sides of the middle line.

The long bones are generally cylindrical or prismatic; the shaft of a long bone is called dyaphysis; its extremities, which are generally enlarged for the purposes of articulating with other bones, epiphyses.

The eminences and projections on the surfaces of bones are called apophyses; they are numerous, and serve for the origin and insertion of muscles, and for articular faces.

The small *foramina*, or holes, on the surfaces of bones serve for the transmission of blood-vessels; the largest is mostly about the middle.

Structure of bones.—The density of bones, though always considerable, and greatly exceeding that of any other part of the body, is variable; being different in different bones, and also in the same bone. This has led to the division of their substance into *compact* and *cellular*, the former being external, the latter internal. The cellular structure grows from the inner surface of the compact, and is composed of filaments and small lamina, or plates, radiating in every direction. The cells resulting from this arrangement communicate with each other, and are filled with marrow. This structure increases the strength without increasing the weight, and also diminishes the effect of concussion, as a fall, blow, &c.

Development of bones.—In the development of bones three stages of ossification are apparent in the embryo. The first is the mucous or pulpy stage, which continues for one month; the second is the cartilaginous, and the third the osseous which commences about the third month. Up to this period the vessels convey lymph only; subsequently they convey red blood to a central point (point of ossification), and there is a deposit of calcareous particles. Most of the bones are formed of several pieces, for each of which there is a distinct point or centre of ossification. These pieces gradually coalesce and form one bone. Bones increase in length by continued deposit at their extremities between the diaphysis and epiphysis, and in thickness by external deposit and by secretion from the periosteum. The latter is proved by the experiment of feeding a young pig on food colored with madder. By suspending and resuming this mode of feeding, alternate lamina of white and colored bone will be produced.

Composition of bones.—All the bones are composed of the same elementary constituents; the relative proportion of these constituents, however, is not always the same. They consist of earthy and animal matters, united generally in the proportion of two parts of earthy to one of animal. Or according to minute chemical analysis they are composed of 32 parts of gelatine, 1 part of insoluble animal matter, 51 parts of phosphate of lime, 11 of carbonate of lime, 2 of fluate of lime, 1 of phosphate of magnesia, and 1 of soda and muriate of soda.

The earthy matter is most abundant in the bones of the head; the animal matter in the cellular structure. In advanced age, and in some forms of disease, the proportion of earthy matter is also increased.

The animal matter may be removed by combustion, and the bone will then be quite white and friable; the original form, however, will be preserved. Immersing a bone in dilute acid will remove the earthy matters, owing to their strong affinity for acids; in this state the animal part remaining is cartilaginous, flexible, and elastic.

Periosteum.—The periosteum is a dense, white, fibrous membrane, which covers the surface of the bones; it adheres with less tenacity in infancy than in adult age. In old age it becomes ossified; it is vascular, and possesses but little sensibility in health; its office is to assist in the secretion of the external laminæ of bone; to protect it against suppuration in the vicinity, restrain the deposit of bone within proper limits, receive the insertion of muscles, tendons, &c.

The delicate vascular membrane, lining the cells, canals and cavities of bone is termed the *internal periosteum.* It contains *marrow,* a substance resembling fat, but of finer consistence. In protracted chronic diseases the marrow is absorbed and its place supplied with serum. It, therefore, would seem to be like fat—a reservoir of nutriment.

Bones are supplied with arteries, veins, lymphatics, and nerves.

Formation of callus.—After a bone is fractured, nature sets to work to repair the injury, and the development of the medium of union, called callus, somewhat resembles the primitive formation of bone. From the vessels ruptured about the seat of the injury blood is at first effused; this is gradually absorbed, and meanwhile coagulating lymph is effused. This coagulates and ossifies in the form of a ring around the broken bone, the thickest part being directly over the seat of the fracture; and also in the form of a pin in the interior of the bone. The extremities of the broken bone now begin to unite or *knit* together, and after this is effected, the superfluous bony matter (the ring and pin) being no longer of any service, are absorbed. They act merely as splints. It is of the utmost importance that the parts be kept at rest whilst this process is going on, otherwise the ossification may be arrested, and a crooked limb, or even a false joint, result.

Bones of the Trunk.

The trunk is formed by the spine, the thorax and the pelvis.

The spine.—The spine extends from the head to the lower part of the pelvis at the posterior part of the trunk. It is formed of 28 or 29 distinct bones called vertebræ, and contains a bony canal for the spinal marrow. It has several curves; in the neck it is convex anteriorly, and concave behind; in the thorax concave anteriorly and convex behind; in the loins convex in front and concave behind; and in the pelvis concave in front and convex behind. Of the 28 or 29 pieces of which it consists, 24 are considered as *true vertebræ* on account of their mobility, and the remainder—5—not being movable, are called *false vertebræ*. They are divided into 7 for the neck, called cervical; 12 to the thorax, called dorsal; and 5 to the loins, called lumbar; the *false* and the sacrum, and 3 or 4 coccygeal bones, situated at the inferior extremity of the column.

A vertebræ consists of a body, 7 processes or extremities, and a canal or foramen for the reception of the spinal cord. The *body* is in front, and in shape is somewhat ovoid; it is convex anteriorly, and concave behind; its surfaces above and below, articulate with a contiguous vertebræ by means of a cartilage, and it is the thickest part.

The *processes*, or extremities, are four *oblique*, which articulate with corresponding ones from the vertebræ above and below; two *transverse*, one projecting on either side from between the oblique processes, and which serve for the attachment of muscles and ligaments; and one *spinous* process, which is placed in the middle of the bone behind, and which also serves for the attachment of muscles.

Cervical vertebræ.—The first of these is called the atlas, on account of its supporting the head; it has no body, but presents the appearance of a large irregular ring, thickest at the sides. In place of a body it is supplied with a bony arch which is occupied by the processus dentatus of the second vertebræ. The *oblique* processes are peculiar, the two above are large, oblong and concave, to suit the condyloid processes of the occipital bone with which they articulate. The two lower ones are round, flat, and horizontal, adapted to the rotary motion of the head.

The second is called *dentata,* from its tooth-like process projecting from the upper surface of the body. The head rotates on this process, which is kept in its place by the transverse ligament of the neck—an exceedingly strong ligament extending across the atlas and attached to a tubercle on each side of that bone. This process is smooth in front where it touches the arch of the atlas, and likewise behind where the transverse ligament plays. The spinous process is long and forked.

The remainder of the cervical vertebræ, except the last, do not require special description; their bodies are small, flattened in front, and gradually increase in size as they descend. The *spinous* processes are short, thick, horizontal, and bifed or forked. The *oblique* processes are flat, oval, and short. The *transverse* processes are broad, and have a foramen at their base for the passage of the vertebral artery. The foramen for the spinal cord is very large.

The *spinous* process of the sixth vertebræ is long and terminated by a sharp point. The seventh is the largest and looks like a dorsal vertebræ; its body is larger than that of the others, and its spinous process is the longest of all, and is not bifurcated, but terminates by a rounded tubercle easily felt under the skin. The foramen at the base of the transverse processes is too small to receive the vertebral vessels.

Dorsal vertebræ.—The dorsal vertebræ situated intermediate between the neck and loins, are also inferior in point of size. They are 12 in number. Their bodies are cylindrical, their transverse diameter decreases from the first to the third, and then regularly decreases to the last. They have articular marks on the sides for the heads of the ribs. These marks or pits, except in the first and the eleventh and twelfth, are each formed by two contiguous vertebræ. The *first* has a complete pit in its side for the head of the rib. The *eleventh* and *twelfth* have also complete fossæ or pits for the ribs. The *oblique* processes are vertical, the two upper looking backward, and the two lower forward. The *transverse* processes are long and terminate in an enlarged extremity, which has an articular face in front for the tubercle of the rib. The *spinous* processes are long, triangular, and broad at the base, sharp-pointed, somewhat rough and swollen at the extremities, and overlap each other.

Lumbar vertebræ.—These are five in number. Their bodies are larger than those of the other *true* vertebræ, oval, and have the transverse diameter the longest. The spinal foramen is triangular, and larger than in the dorsal vertebræ. The groves for the nerves to pass out are likewise large. The *oblique* processes are vertical; the superior looking inwards and the inferior looking outwards. The *transverse* processes are very long, and stand out at right angles. The *spinous* processes are short, thick, quadrangular, horizontal, and terminate by an oblong tubercle. The lumbar vertebræ increase in size successively—the first is the smallest, and the fifth the largest; the body of the latter is also wedge-shaped. The third has the longest transverse and spinous processes.

Sacrum.—The os sacrum is much the largest bone in the spinal column. It forms the upper and posterior boundary of the pelvis; it is triangular in shape, with the base above, and originally consist of five bones, bearing a strong resemblance to vertebræ; hence they received the name of false vertebræ; The anterior face of this bone is smooth and concave, and has four holes on each side, which communicate with the spinal cavity and transmit the anterior branches of the sacral nerves. Its posterior face is convex and rough, and is equally divided by its spinous processes. The marks of the original separation of this bone are evident, particularly on its front surface. The base has an articular mark to receive the last lumbar vertebræ: the *apex* is below and has an articular surface for the coccyx. The *sacral* canal, a continuation of the spinal, runs through the length of the sacrum. The portion of the spinal cord lodged within it is called the cauda equina, from its fancied resemblance to a horse's tail.

Coccyx.—The coccyx, like the sacrum, is triangular; it has its base upwards, and is flat. It consists of three or four bones united in the same curve as that of the sacrum. These pieces as life advances become joined together by bone, and also joined to the sacrum; so that all the false vertebræ, from the base of the sacrum to the point of the coccyx, are united into a single bone. The coccyx forms the lower extremity of the spinal column, and corresponds with the tails of animals.

The vertebral column affords a secure lodgment to the spinal marrow, is a line of support to the trunk in every position, and is the centre of all its movements. In the erect posture, the spine also supports the head.

The motions of the spine are, flexion or bending forward, extension or bending backward, lateral bending and circumduction, or that motion in which the trunk is caused to describe a cone, the base of which is above, and the apex below.

The first of these motions, flexion, is the most extensive; it is produced mainly by the action of the abdominal muscles. In this position the inter-vertebral, elastic, cartilages are compressed in front, and thickened behind, the anterior vertebral ligament is relaxed, and the posterior vertebral ligament correspondingly tense.

Extension or backward bending is very limited, owing to the mechanical obstruction offered by the spinous processes of the back and neck, which are very near to and overlap each other. The abdominal muscles also strongly resist it. The muscles, arising either from the back of the pelvis or from the transverse processes, and going upwards to be inserted either into the ribs, the transverse or spinous processes, which produce this motion, act less advantageously than do the abdominal muscles in producing flexion; the leverage in the latter case being much more favorable.

The lateral inclination of the spine is considerable, it is favored by the advantageous position of the muscles on the side of the neck and trunk, as well as by the mechanical arrangement of the parts; the transverse processes being so far apart as scarcely to offer any resistance. The principal obstruction is presented by the ribs, which strike against each other when this motion is carried too far.

Circumduction is performed chiefly on the lower dorsal and the lumbar vertebræ, and is a succession of the movements above described.

The rotation of the spine is very limited; it is performed chiefly on the lower dorsal and upper lumbar vertebræ, and consists of a series of oblique motions of the body of one vertebræ upon another. In old age, when the inter-vertebral substance has become hardened, it is almost inappreciable.

The rotation of the head is effected by the motion of the atlas upon the dentata, which is the only motion allowed to this vertebræ, flexion and rotation being prevented by the confinement of the transverse ligament behind, and by the anterior bridge of bone in front. The flatness of the articular surfaces of the two bones likewise prevent any lateral motion.

Pelvis.—The bones of the pelvis are the sacrum, the coccyx, and the two innominata or hip bones; the two first, which form the posterior boundary, have already been described.

Innominatum.—The innominatum is a large flat bone, and is the hip or haunch bone of common language; it has some resemblance to the figure 8. From having originally consisted of three pieces, though these pieces unite in the cavity of the hip joint (acetabulum), and in the adult leave but a slight trace of their former distinction, anatomists divide the innominatum into the ilium, ischium, and pubes.

Ilium.—This is the largest of the three pieces, and forms the upper rounded part of the innominatum, or the wall of the upper pelvis. It is the hip bone. Externally its surface is convex and rough, with a semi-circular ridge crossing it; the glutei muscles arise from this surface. Internally it is concave; the anterior portion is smooth and gives origin to the iliacus internus muscle; the posterior is rough and articulates with the sacrum, behind which arise muscles and ligaments. The margin, called the crista or spine of the ilium, is curved and somewhat resembles the italic *f*. In front of the crista there are two eminences or projections, one of which is called the anterior superior spinous process, and the other the anterior inferior spinous process; the former gives origin to the sartorius and tensor vaginæ muscles and Poupart's ligament; the latter to the rectus muscle; the cavity between the two gives origin to the gluteus medius. A large prominence called ilio-pectineal is situated below these processes; in a groove above this, the iliacus internus and psoas magnus pass.

The posterior margin of the ilium is marked by two projections called posterior superior, and posterior inferior spinous processes. Just below the latter there is a deep excavation called sciatic notch, through which the pyriform muscle, the sciatic nerve, and several blood-vessels pass out. The *crista* has three lips, the internal of which gives origin to the transversalis muscle, the middle to the internal oblique, and into the external the external oblique is inserted.

Ischium.—This portion of the innominatum is the most inferior and is the next in size to the ilium; it consists of a body and ramus or branch. The external surface of the body is rough; the internal surface is smooth, and is called the plane of the ischium. On the posterior margin there is a projection called the spine, into which the lesser sacro-sciatic ligament is inserted; beneath the spine there is a groove in which the tendon of the obturator internus muscle plays. The inferior portion of the body is called the tuberosity of the ischium; it gives origin to the biceps adductor, semi-membranosus and semi-tendinosus muscles; in front there is a long ridge into which the great sacro-sciatic ligament is inserted. The ramus is short and thick, and ascends forward and inward to join the ramus of the pubes, and form a part of the pubic arch; it is rough externally, and smooth internally; the crus of the penis arises from the latter face.

Pubis.—This bone is much the smallest of the three, and constitutes the anterior boundary of the pelvis. It consists of a body and two branches, one of which runs downwards to join the ischium, and the other backwards and upwards to join the ilium. The body is joined to its fellow by a flat surface, called the symphysis; the superior portion of the body is horizontal at right angles with the symphysis, and is bounded outwardly by a projection, called the spinous process. From this process two ridges proceed outwardly; the posterior, is called the crista of the pubes, or linea pectinea, and to it is attached a part of Poupart's ligament. Between these two ridges is a triangular space from which the pectineus muscle arises, and over which the femoral vessels pass. The extremity of the superior ramus is triangular, and much enlarged where it contributes to form the *acetabulum*, a hemispherical concavity for the articulation of the femur. At the bottom of this cavity there is a rough depression occupied by a mass of fat generally called a gland of Havers. The obturator or thyroid foramen is the large opening in the lower part of the bone. In shape it is mostly oval; it is filled up, with the exception of a groove in the upper part to transmit the obturator vessels and nerve, by a membranous ligament.

6

As a whole, the pelvis is a conoidal cavity, having its base above, and its apex below. Internally it forms an irregular floor for the support, in the erect position, of the abdominal viscera. Externally its projections furnish favorable points for the attachment of muscles. There are marked differences in the male and female pelvis. The diameters of the latter are larger, and the depth less than in the male. The, arch as it is called, under the symphysis pubis, is regularly rounded in females; in males, on the contrary, it is merely an acute angle. The sacrum also, in women is shorter, more concave, and of greater breadth.

Thorax.

The thorax is the upper part of the trunk, and in shape is conoidal, flattened in front, and concave behind; it is formed by the dorsal vertebræ behind, by the sternum in front, and at the intermediate spaces by the ribs.

The superior part, or apex, of the cone presents a heart-shaped opening, and is much smaller than the base; the latter has a large notch in front, into which the lower end of the sternum projects.

Fig. 1.

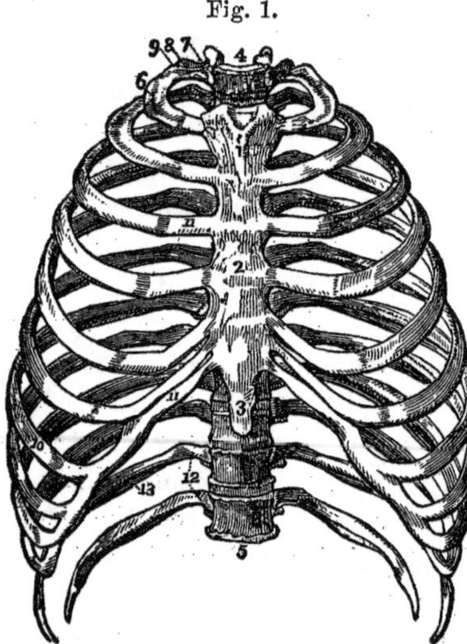

Fig. 1. A representation of the thorax; 1, 2, 3 bones of the sternum; 4, 5 first and last dorsal vertebræ; 6, 10 angle of ribs; 11 cartilages.

Sternum.—This bone is placed in the middle and front of the thorax, and is oblong and slightly curved; its direction is also slanting, the lower part receding much farther from the spine than the upper. The concavity is in front. In the adult it is usually composed of 3 pieces; an upper, middle, and lower, which are held together by cartilages and ligaments. Frequently in advanced life these pieces are fused together and form but one bone.

The *first* or upper bone is the thickest, and has an irregular square figure; the upper margin projects somewhat, and is slightly scooped out. On each side are marks for articulating with the clavicle, and first and second ribs.

The *second* or middle bone is longer and narrower than the first; the lower part increases somewhat in breadth, and then terminates by being rounded off on either side. The sides of this piece present pits for the articulation of part of the second, the third, fourth, fifth, sixth, and part of the seventh ribs.

The *third* bone of the sternum in young persons is very often wholly cartilaginous, and is called the zyphoid or ensiform (sword-shaped) cartilage. It is thin, and varies very materially in shape in different individuals. The extremity is mostly pointed, and inclined slightly forwards; sometimes it is forked, and sometimes also turned backwards.

Ribs.—There are twenty-four ribs in all, twelve on either side; the upper seven, from their cartilages articulating with the sternum are called the sternal or *true* ribs; the lower five in consequence of their cartilages stopping short of the sternum are called *false* ribs.

In some rare cases there have been eleven or thirteen ribs on a side.

The ribs are parallel, and directed obliquely downwards and forwards; each having a parabolic curve, and gradually increasing in size to the eighth, and afterwards gradually diminishing. The surface of each rib is convex externally, and concave internally; the anterior or sternal extremity is larger and flatter than the posterior; the vertebral or posterior extremity of the rib is its head, and presents two articular surfaces separated by a ridge. Surrounding the head the rib is very narrow, and called the neck. Near the vertebral extremity is an oblique ridge caused by the insertion of the sacro-lumbalis muscle; just at this line a curve, somewhat abrupt, takes place, which is the angle of the rib.

The upper margins of the ribs are rounded, and somewhat

rough for the insertion of the intercostal muscles; the inferior margins have a thin cutting edge, just with and above which is a groove running two-thirds of their length, which contains the intercostal vessels and nerve. Each rib is twisted and bent, the spinal extremity being directed upwards, and the sternal extremity downwards. About an inch from the head of each rib is a tubercle, or prominence, for articulating with the transverse processes of the vertebræ; just beyond this is a smaller tubercle for the insertion of the external transverse ligament.

The *first* rib is small, and more circular than the other; it is flat above and below.

The *eleventh* and *twelfth* are not connected with the others, and are thence called floating ribs; they have no tubercles, and but one articular surface on the head.

The thorax performs two important offices in the human economy—first it protects the organs of circulation and respiration, and second, it assists in the function of respiration. Its structure affords a very secure defence to its viscera, from the effect of blows on the outside. Posteriorly the spinal column, and the masses of longitudinal muscles which fill up the gutters on each side of the spinous processes furnish, ample protection. In front the protection is less secure, in consequence of the sternum being placed immediately under the skin. The effects of blows, however, on this part are much diminished by the elasticity of the cartilages, and by the oblique downward direction of the ribs themselves; both of which circumstances dispose the sternum to retreat backward, and to yield to the impelling force. The arched form of the thorax is likewise a favorable one to enable it to resist the force of blows, &c.

The thorax aids respiration by expanding and contracting; the expansion takes place in three directions, vertically, transversely, and antero-posteriorly, or from back to front. In the vertical direction the augmentation is effected by the diaphragm; it is much more considerable in adults than in children, owing to the viscera of the abdomen being comparatively larger in the latter. In the other directions the dilatation is accomplished chiefly by the action of the intercostal muscles, which contract successively—beginning at the first rib which serves for a fixed point—and elevate the ribs. The obliquity of the ribs and the attachments of their cartilages to the sternum, which have to ascend in order to reach it, allow of this augmentation. While the enlargement is going on the sternum is slightly elevated and projected forwards.

In expiration all the diameters of the thorax are diminished, all its movements being directly the reverse of those which take place in inspiration. The chief cause of expiration is atmospheric pressure upon the external walls of the thorax, acting in conjunction with the natural elasticity of the lungs. Some slight aid is also furnished by muscular contraction and the elasticity of the cartilages.

Bones of the Head.

In the head there are twenty-two bones, which are divided into those of the cranium—eight in number,—and those of the face—fourteen in number.

Cranium.

The cranium is composed of one frontal, two parietal, two temporal, one sphenoid, and one ethmoid bone.

The frontal bone is situated in front of the cranium; the occipital bone is at its hind part; the two parietal bones, one on each side, form its superior lateral part; the two temporal also, one on each side, its inferior lateral part; the sphenoid is situated at the middle of its base; and the ethmoid also at its base, and just in front of the body of the latter bone.

Frontal Bone.—This bone forms the forehead of common language; it is symmetrical and shell-like, and is occasionally divided into two parts by a suture, continuous with that, dividing the parietal bones. The *external* surface is convex, and the internal concave. About the middle of the external surface on each side is a protuberance called the frontal, which is the original centre of ossification. The *internal* surface is concave, and marked by numerous depressions, corresponding with the convolutions of the brain. Near the middle and lower margin of the external surface is a ridge called the nasal, or superciliary protuberance; just below this there is a protuberance called the nasal spine, which serves for an abutment to the nasal bones. At the inferior margin of the bone on either side are the two orbital ridges, which form the anterior boundary of the eye. These ridges are terminated outwardly by the external angular processes, and inwardly by the internal angular processes. The lachrymal gland is placed in a depression just within the external angular process.

The frontal sinuses consist of one or more large cells placed beneath the nasal protuberances; they vary very much in size, and in some rare instances in the adult do not exist.

This bone unites laterally with the parietal and sphenoid; and below with the ethmoid and several of the bones of the face.

Parietal Bones.—These, as stated, occur in pairs and form the middle, superior and lateral portions of the cranium; they are quadrangular, convex externally, and concave internally. About the middle of the external surface the bone is raised into the parietal protuberance—the centre of ossification. Immediately below this is a semi-circular ridge for the attachment of the temporal fascia and muscle. *Internally*, the surface of the bone is marked by the convolutions of the brain, and contains also a number of furrows, produced by the middle artery of the dura mater. The superior margin is the thickest, and much serrated; when joined to its fellow it forms a deep groove for the accommodation of the longitudinal sinus. The inferior edge is thinner and arched. This bone articulates with its fellow, with the frontal, the sphenoid, the temporal and the occipital bones.

Occipital Bone.—This bone is placed at the posterior and inferior part of the head, and is of an oval or trapezoidal shape; it is convex externally and concave internally. Both surfaces contain a number of ridges and processes. A large opening, called the foramen magnum, is found in its lower part. This hole is oval, the long diameter extending from before backwards;

it transmits the medulla oblongata, the vertebral vessels, and the spinal accessory nerves. On each side is an oblong convex surface, called condyloid process, for articulating with the atlas. Near the middle of the external surface of the bone is a prominence, called the external occipital cross, from which, on each side, there proceeds a semi-circular ridge, which serves for the origin and insertion of muscles. About an inch below this ridge is another, into which the superior oblique muscle is inserted. In front of the foramen magnum is the basilar process, the extremity of which joins the sphenoid bone. The internal surface is likewise marked by a cross near its centre, called the internal occipital cross, which is more prominent than the external; from this cross three groves diverge, which contain the two lateral and the superior longitudinal sinus; inferiorly a ridge proceeds to the foramen magnum to which the falx cerebelli is attached. In this way four concavities are formed, the two superior of which contain the posterior lobes of the cerebrum, and the two inferior the hemispheres of the cerebellum.

Superiorly the occipital bone joins the parietal, laterally the temporal, and in front the sphenoid.

Temporal Bones.—These bones are placed on either side of the cranium, below the parietal, and form portions of its inferior lateral parieties and base. They are divided into three portions, the squamous, the mastoid, and the petrous.

The *squamous* portion is thinner than the other bones of the cranium, but is covered by the temporal muscle, which affords sufficient protection to the brain. Externally it is slightly convex, and has grooves for the deep temporal artery. The zygomatic process projects from its lower part, and forms a part of the zygomatic arch. Beneath the base of this process, which is triangular, is the glenoid cavity for the articulation of the lower jaw. Just behind this cavity is another, which contains a portion of the parotid gland. Between these cavities is the Glasserian fissure, which transmits the chorda tympani nerve, and the levator tympani muscle, to the ear. The internal surface has a groove for the middle artery of the dura mater.

The *mastoid* portion is behind, and is thick and cellular; its upper part is angular, has serrated edges, and is received between the parietal and occipital bones. Below there is a large conical projection, called mastoid process, which receives the insertion of sterno-mastoid and trachelo-mastoid muscles. At the base of this process is a fossa from which the digastric muscle arises. A foramen, called the mastoid, and which transmits a blood-vessel, is found near the upper margin of the bone.

There are numerous and large cells in this portion of the bone, which are called sinuses; they communicate with the tympanum by a large orifice.

The *petrous* portion is triangular and pyramidal; it arises by a broad base from the inner portions of the mastoid and squamous portions; its structure is extremely dense and brittle. In the posterior surface of the petrous portion is a large foramen, called the meatus auditorius internus, through which the seventh or auditory, and the facial nerves are transmitted. The base has a large oval opening, between the zygomatic and mastoid processes, which leads to the tympanum, and is called the meatus auditorius externus; its margin is called the auditory process, its lower part is rough for the attachment of the cartilage of the external ear. In the angle between the squamous and petrous portions is the orifice of the Eustachian tube. The styloid process, which is round, tapering, and about an inch and a half long, projects from a ridge on the inferior surface, called the vaginal process. Immediately behind the base of the styloid process is the stylo-mastoid foramen, through which the facial nerve and stylo-mastoid artery pass. The internal jugular vein, and the eighth pair of nerves pass through the posterior foramen lacerum, which is formed by the articulation of this with the occipital bone. Just before the foramen lacerum is the opening of the carotid canal, which contains besides the artery the ganglion of Laumonier.

The temporal bone articulates with the occipital, the parietal, the sphenoid, and the malar.

Sphenoid Bone. — This bone is placed transversely in the

middle of the base of the cranium. In shape it bears resemblance to a bat; it consists of a body and four wings, two, a large and a small one, being placed on each side. The body is in the centre, and is cuboidal in shape. Besides the wings mentioned it has a number of angular margins, and additional processes, two of the latter are directed vertically downwards. A deep depression exists on the superior surface of the body, which is called the sella turcica, and contains the pituitary gland. On either side of this depression are two grooves, called sulci carotica, for the carotid arteries; in front is a prominence, called processes olivaris, upon which is a groove, indicating the position of the chiasm of the optic nerves. The posterior clinoid process projects over the depression behind. In front of the body is a ridge, which articulates with the nasal lamella of the ethmoid bone, on either side of this are the orifices of the two sphenoidal cells; these empty into the posterior ethmoidal cells, and do not exist in infancy. The posterior surface of the body is flat, and rough for articulating with the cuneiform process of the occipital bone. Generally in the adult the bones are anchylosed at this point.

The small wings are in front of the large, and are triangular, flat, and narrow. Their posterior extremities form the anterior clinoid processes, and are perforated by the optic foramen, which transmits the optic nerve and ophthalmic artery. They articulate with the frontal bone.

The great wings are divided from the small, by the sphenoidal fissure or foramen, through which pass the third, fourth, first branch of the fifth, and the sixth nerves; it has three surfaces. One is anterior, and called orbital, from its forming a part of the orbit; another is external and called temporal; the third, towards the brain, is concave, and contains a considerable portion of the middle lobe of the cerebrum; its face is marked by the convolutions of the brain, and by a furrow for the passage of the middle artery of the dura mater to the cranium. The temporal surface is somewhat concave, and covered by the temporal and external pterygoid muscles. The orbital face is square and slightly concave. The inferior portion of the great wing is elongated backwards, and called the spinous process. From the point of this process the styloid process projects downwards.

The sphenoid bone articulates above and in front with the vomer, the frontal, ethmoid, malar, and parietal bones; behind with the occipital, and by its pterygoid processes with the palate bones; and laterally with the temporal.

Ethmoid Bone.—The ethmoid bone, so named from its resemblance to a sieve, is placed at the base of the skull and between the orbitar processes of the frontal. In shape it is cuboidal, and in structure extremely cellular and light. The superior surface is called the cribriform plate, and is perforated with holes which transmit filaments of the olfactory nerve. From the middle line of this surface, a narrow, triangular, hollow process projects called the crista galli, to which is attached the falx cerebri. A vertical plate of bone, called the nasal lamella, divides the bone longitudinally into two halves; it articulates posteriorly with the crista sphenoidalis, and below with the vomer.

The lateral surfaces of the bone are called ossa plana; they are extremely thin and form a large portion of the orbits of the eyes. Attached to the internal face of the os planum on either side of the nasal lamella are two scrolls or shells placed one above the other, and called superior and middle turbinated bones. They are separated by a fissure which is the superior meatus of the nose. The middle meatus is between the middle and inferior turbinated bones. The anterior ethmoidal cells empty into the middle meatus; the most anterior of them is funnel shaped and receives the fluid from the frontal sinus. The posterior ethmoidal cells and the sphenoidal sinus empty into the superior meatus.

The ethmoid articulates with the frontal, sphenoidal, inferior turbinate, superior maxillary, nasal, lachrymal, palate bones, and vomer.

Bones of the Face.

The face is composed of fourteen bones viz. the superior maxillary, the palate bone, malar, nasal, unguiform, inferior turbinated, vomer, and inferior maxillary; all of these except two, the inferior maxillary and vomer, are in pairs, there being six on either side. The inferior maxillary bone has corresponding or symmetrical sides.

Superior Maxillary Bones.—These constitute the upper jaw and are the largest bones of the face; they articulate with all the others except the lower jaw. They have an irregular cuboidal body, and four processes. The body is hollow, and has four surfaces. The anterior or facial surface is bounded above by the lower margin of the orbit, below which is a foramen, the infra orbitar, which transmits the infra orbitar nerve and an artery. The posterior or temporal surface is round; it has a roughened prominent part, called the tuber, through which pass by several small foramina, the posterior dental nerve, artery, and vein to the floor of the antrum. The superior or orbital surface is triangular. The internal or nasal surface has a large opening of the antrum highmorianum—the large cavity in the centre of the bone. The walls of this cavity are grooved by the passage of the dental nerves.

The nasal process arises by a broad and thin base from the upper and anterior part of the bone; it articulates anteriorly with the nasal bone, and superiorly with the frontal; the posterior edge forms a canal, by articulating with the os ungus, to contain the lachrymal sac.

The malar process is on the upper and outer surface of the bone; it is rough and articulates with the malar bone.

The alveolar process, contains the sockets of the teeth, eight to each side; it is broader behind than in front. Just behind the alveolar process, at the junction of the two bones, is the foramen incisivum which contains the naso-palatine nerve and ganglion.

The palate process arises from the internal face of the body of the bone, just within the circle of the alveoli; it forms the horizontal roof of the mouth, and floor of the nose. The nasal crista arises at the junction with its fellow and articulates with the vomer. Behind, this process unites with the horizontal part of the palate bone; its anterior extremity forms the anterior nasal spine.

Palate Bones.—These bones, two in number, are placed between the superior maxillary and the sphenoid bones. They are irregular in shape, and divided into two portions.

The horizontal, or palate plate, is square, and in the same line with the palate process of the superior maxillary bone, forming a part of the floor of the nostril and roof of the mouth. At its suture it forms part of the nasal crista for articulating with the vomer. Behind, it is elongated into the posterior nasal spine.

The vertical or nasal plate forms the posterior and outer part of the nostril, and is much thinner than the palate plate; its nasal surface articulates by a ridge with the inferior turbinated bone; externally it articulates by a roughened surface with the maxillary bone. Posteriorly it has an elongated triangular process, called pterygoid.

The orbital portion, or plate, is hollow and irregular; it forms a small part of the orbit between the ethmoid and superior maxillary bones.

The palate bone articulates with the upper maxillary, the sphenoid, ethmoid, inferior spongy, the vomer, and with its fellow.

Nasal Bones.—These are placed between the nasal processes of the superior maxillary bones; they are oblong in shape, and compact in structure, and so placed together, as to form a strong arch, called the bridge of nose. The upper extremity is thick and narrow, and articulates with the frontal bone. The lower is thinner and broader, and has the cartilages of the nose attached. They articulate with each other, with the nasal processes of the superior maxillary, and with the frontal bone.

Unguiform Bones.—So called from its resemblance to a finger nail, likewise called lachrymal. It is placed at the internal side of the orbit, between the nasal process of the upper maxillary, and the os planum of the ethmoid, and is very small and thin. The external surface forms a part of the orbit of the eye. It articulates loosely with the frontal, upper maxillary, os planum, and the inferior spongy bone.

Malar Bones.—The malar or cheek bones, two in number, are quadrangular in shape. They are situated at the external part of the orbit of the eye, and form the prominence of the cheek. The cheek bone has three surfaces; the external or facial, which is convex, and forms part of the face; the internal or orbital, which is crescentic, and contributes to the orbit; and the posterior, which is concave, and forms the anterior boundary of the zygomatic fossa. It has four margins, two of which are superior, and two inferior. The anterior of the upper is curved to form the external and part of the inferior margin of the orbit; the posterior upper is irregular, and has attached to it the temporal fascia. The anterior inferior is serrated, and articulates with the upper maxillary; the posterior inferior serves for the origin of a part of the masseter muscle. The angles of this bone are called processes, the superior or frontal articulates with the angular process of the frontal bone. The temporal process extends backwards, and forms part of the zygomatic arch; the maxillary articulates with the upper maxillary bone.

The malar bone articulates with the maxillary, frontal, sphenoidal, and temporal bones.

Inferior turbinated Bones.—These, also, called spongy bones, are situated immediately below the opening into the antrum Highmorianum. They are scroll-like, and extremely light and porous. The internal surface is convex, and looks towards the nose; the external surface has a broad hook which rests upon the lower margin of the orifice of the antrum and partly closes it. Two processes arise from the superior margin of the bone, and join the ethmoid. The lower margin is thicker than the upper.

Vomer.—Called so from its resemblance to a ploughshare. It is a single bone placed between the nostrils, and forms a large part of the septum. It is formed of two plates of compact structure. It is flat, has four margins, and is frequently more inclined to one side than the other. The superior margin is the thickest and contains a groove for the reception of the azygos process of the sphenoid; the inferior margin articulates with the nasal crista of the upper maxillary and palate bones; the anterior margin is directed obliquely downwards and forwards, its front half joins the cartilaginous septum of the nose, and its posterior half unites with the nasal lamella of the ethmoid; the posterior margin is concave, sharp, and thin, and divides the posterior opening of the nose.

Inferior Maxillary.—This bone, commonly called the lower jaw, is also single. It forms the lower boundary of the face, and is the only bone in the head capable of motion. Its shape is that of a parabolic curve. In infancy its halves are separable, being joined in the middle only by cartilage; two or three years after birth, however, they become consolidated, and the cartilage disappears. It consist of a body or region which corresponds with the teeth, and two rami or branches.

The body is convex in front, having its upper part formed by the alveolar cavities for receiving the teeth, and its lower part presenting a thick and rounded margin, which is the base of lower jaw. At the middle or symphysis is a triangular prominence, called the anterior mental tubercle. Near to this tubercle, on either side, there is a large hole, called the anterior mental foramen, which conducts the inferior maxillary artery and nerve to the teeth.

The internal or posterior face is concave, and has also a prominence at its symphysis, called the posterior mental tubercle; a fossa exists on each side of this tubercle for the insertion of the digastric muscle. Exterior to this on either side is a larger fossa for the sublingual gland.

The alveolar processes are thin with cutting edges; and contain sockets for sixteen teeth. They come and go with the teeth; it is the absorption of these processes, after the loss of the teeth, in aged persons, and the consequent falling in of the mouth that gives to the chin its prominent appearance.

The ramus or extremity of the lower jaw is square, and elevated above the body. In middle age it is at right angles to the latter, and in youth and old age oblique. The superior margin has a thin, concave edge, and is bounded in front by the coronoid, and behind by the condyloid process. The first of these processes is triangular and thin, and receives at its point the insertion of the temporal muscle; the second articulates with the glenoid cavity of the temporal bone, and is attached to the ramus by a narrow neck on the inside of which is a fossa for the insertion of the external pterygoid muscle. Externally the surface of the ramus is rough; internally there is a foramen for conducting the inferior alveolar artery and nerve to the teeth. The junction of the body and ramus is the angle of the jaw; it is rough for the attachment of the stylo-maxillary ligament.

The lower jaw has a greater influence on the form of the face than any of the other bones. In some persons it is much smaller in proportion than in others. The inclination of the alveolar processes externally or internally, frequently cause the chin to recede or project unusually. At times, also, the separation of its sides, shorten the chin, and give increased width to the lower part of the face.

General Remarks on the Head.

In the adult the bones of the head consist of two tables, one external, and the other internal. The external table is hard and fibrous, the internal thinner and more brittle, and hence has received the name of vitreous table.

These tables are united by a cellular bony substance, called *diplœ*, which is traversed by channels, filled with veins having delicate parietes, and provided with valves, and which like all other veins are intended to return the blood to the heart.

They empty at the base of the brain into the emissaries of Santorini.

The bones of the cranium are united by seams or *sutures* which, except in old age, are very distinctly marked. They are formed by the junction of the edges of the bones, which are serrated or notched, and accurately and firmly fitted into each other. Indeed, in some of the sutures uniting the flat bones of the head, a complete dove-tailing is, here and there, met with.

This kind of union is often confined to the external table and to the diploe, whilst in the internal table the bones are united by a nearly straight joint.

The principal *sutures* of the head are: the *coronal*, so named from its corresponding in situation with the garlands, worn by the ancients, which unites the parietal and frontal bones; the *saggital*, which unites the *two parietal* bones in the adult, and in the child extends through the frontal bone to the root of the nose; the *occipital*, or lambdoidal, uniting the parietal and occipital bones, and also the latter to the temporal bones—the latter half is sometimes called the additamentum suturæ lambdoides; and the squamous which joins the parietal and temporal bones on either side of the head. One or more small bones are often met with in the upper part of the lambdoidal suture, attached to the parietal and occipital bones by serrated margins; they are called *Ossa Wormiana*, or *Triquetra*.

Besides these sutures, which belong exclusively to the head, there are others uniting the bones of the face.

Much speculation has been entered into in regard to the *uses* of the sutures; they appear, however, to be only a provision for the growing state, as they cease when this terminates. In the fœtus they are often serviceable in facilitating delivery, by permitting the head of the child to accommodate itself to the pelvis of the mother.

The *Fontanelles.*—These are membranous spaces peculiar to the fœtus and to early infancy. They are owing to a deficiency of bone at their angles of junction.

In the bones of the cranium the points of ossification beginning at the centre, and going towards the circumference, the angles, being the longest radii, are last in ossifying. Hence they are generally incomplete at birth. The fontanelles are six in number, two on the middle line of the head above, and two on either side. The *anterior*, known in common language as the "opening of the head", is the largest. It is quadrangular or lozenge shaped, the anterior angle generally being the longest, and is situated at the fore part of the sagittal suture. The *posterior* fontanel is situated at the junction of the lambdoidal and sagittal sutures; it is small and triangular, of the two fontanelles on either side, one is at the angle of junction of the temporal, occipital, and parietal bones; the other at the junction of the parietal and sphenoid bones. They are quite small and often indistinct. All of these fontanelles ossify speedily after birth, and are often entirely closed by the end of the first year.

Nasal cavities.—These cavities are large and irregular; they are divided by the nasal septum. Each nostril or cavity has three distinct passages, or *meatuses,* as they are sometimes called —a superior, a middle, and an inferior. The superior meatus is bounded by the superior and middle turbinated bones; the posterior ethmoidal and sphenoidal cells, and the spheno-palatine foramen open into it. The middle is between the middle and lower turbinated bones, and has opening into it, the anterior ethmoidal cells, the frontal sinus, and generally the antrum. The inferior is bounded above by the inferior turbinated bone, and below by the floor of the nostril; it is the largest, and at its upper and anterior part is the nasal duct which communicates with the orbit of the eye. The opening of the nostrils in front is called anterior nares; the opening behind posterior nares. The mucous membrane lining the nostrils is called the Schneiderian membrane; it extends into the various cavities which empty into the nostrils, and when in a state of irritation from cold, &c., the secretions from it are often copious and exceedingly annoying.

Orbits of the Eyes.—The orbital cavities are conical, having the base outwards, and the apex backwards; they are also somewhat quadrangular. Seven bones unite in forming each orbit namely, the frontal, malar, superior maxillary, ethmoid, unguis, sphenoid, and palate bones.

The roof or superior face of the orbit is formed by the frontal and the lesser wing of the sphenoid; the floor or inferior face by the upper maxillary, malar, and palate bones; the external face by the malar and greater wing of the sphenoid; the internal face by the unguis, or lachrymal, os planum of the ethmoid, and the body of the sphenoid.

The apex of the orbital cavity contains the optic foramen for the transmission of the optic nerve. Several other foramina for the passage of blood-vessels and nerves also open into the orbits.

The *Facial Angle.*—The bones of the face usually project beyond those of the cranium. In general the races of man that are most perfectly developed have this projection less than those of inferior conformation. The facial angle has been instituted to denote the differences in this respect. It is formed by two straight lines,—one drawn from the lower part of the frontal bone, to the anterior nasal spine, at the orifice of the nose between the roots of the middle incisor teeth of the upper jaw,—and the other from this latter point through the external meatus of the ear. The angle included between these lines has been found in Caucasian or European heads to be about eighty degrees; in the heads of the Mongolian or copper-colored races about seventy-five degrees; and in those of the Negro or Ethiopian about seventy degrees. The nearer the angle approaches to a right angle the greater is the degree of intellectual cultivation manifested. Even in Europeans, however, this rule is liable to many exceptions, the facial angle varying in its size from causes which have no connection with the development of the brain.

The *internal* surface of the cranium is regularly arched above; below it presents three deep fossæ on each side. The anterior fossæ contains the anterior lobes of the brain, the middle the middle lobes, and the posterior is for the cerebellum. The entire surface is marked by round superficial depressions made by the convolutions of the brain, and is lined by the dura mater. In adults the diameters of the cavity of the cranium are, about six and a half inches from front to rear, five inches laterally, and five vertically.

Hyoid Bone.

This is an insulated bone placed in the neck at the root of the tongue. In shape it resembles the letter *u*, and has the convexity in front. It consists of a body and four cornua or horns; the body is in the middle, is convex in front, concave behind, and is the largest part of the bone; its front gives origin and insertion to muscles; two of the cornua are called the greater, and two the lesser, the former are about an inch and a half long, and mostly united to the body by cartilage and ligaments; they are flattened, project backwards, and terminate in a tubercle. The lesser cornua are small cartilaginous bodies, three or four lines long, placed at the junction of the body and great cornua.

They are frequently found ossified; a ligament, the stylo-hyoid, is attached to each.

Upper Extremities,

This part of the skeleton is divided on either side into the shoulder, arm, fore-arm, and hand. Two bones belong to the shoulder—the scapula and the clavicle.

Scapula.—This bone, known as the shoulder-blade, in ordinary language, is placed on the back part of the thorax, and extends from the second to the seventh rib. Its shape is triangular, and it is also thin and flat, presenting two surfaces—one anterior, and one posterior; three edges or margins—one being superior, another external, and the third internal, or posterior, the latter being parallel with and near to the spinal column;—and three angles, one of which is superior, another inferior, and the third exterior or anterior.

The posterior surface or dorsum is somewhat concave, and is divided by the spine into two fossæ, of which the lower is much larger than the upper; the latter, called fossa supra spinata, is occupied by the supra-spinatus muscle, and the former the fossa infra spinata, by the infra-spinatus muscle. The spine is a rough process beginning at the posterior edge of the bone, and running obliquely across it towards the anterior angle, and terminating in the acromion process, which is flat and triangular, overhangs the shoulder joint, and has a small articular mark in front for the clavicle. The trapezius muscle is inserted into, and the deltoid arises from, the edge of the spine.

The internal edge, that parallel to the spine, is the longest, and into it are inserted the rhomboid and serratus anticus muscle.

The superior edge is thin and small, and has a notch in it, called the coracoid notch, through which the scapular artery and nerve pass.

The external edge is thick, and gives origin to the teres minor, and the long head of the triceps muscle.

The anterior angle is converted into a large shallow cavity, called glenoid, for articulating with the head of the humerus. The long head of the biceps arises from a flat surface at its upper end. Immediately blow the glenoid cavity the bone is narrowed and thickened; this part is termed the neck, or cervix, and from it a curved process projects forwards and outwards called the coracoid. This process serves for the attachment of muscles and ligaments.

Clavicle.

The clavicle is a long bone placed transversely at the upper and anterior part of the chest. It articulates with the sternum and the acromion process of the scapula, and in shape is curved somewhat resembling the italic letter *s*. The sternal two-thirds of the bone is convex, and the outer third concave anteriorly; it is cylindrical in the middle, flattened at the humeral, and triangular at the sternal extremity. In males it is shorter, thicker, and more crooked than in females.

The superior face is smooth, and marked near the sternum with a depression for the origin of the sterno-cleido mastoid muscle. The inferior face is roughened near the sternum for the costo-clavicular or rhomboid ligament; near the outer extremity is a rough tubercle for the attachment of the coraco-clavicular ligament; between the ends a superficial fossa exists, for the subclavius muscle. From the sternal two-thirds of the *anterior* edge the pectoralis major muscle arises, and from the humeral third of the same edge the deltoid has its origin. Near the middle of the *posterior* edge one or more foramina are presented for nutritious vessels.

Humerus.

The humerus, or arm-bone, is cylindrical, and extends from the shoulder to the elbow. It consists of a superior extremity, also called the head, a neck, an inferior extremity, and a body.

Fig. 2. The humerus. 1 the body; 2 the head; 3 the neck; 10 nutritious foramen; 13 external condyle; 14 internal condyle; 11 articulating face for the radius; 12 articular face for the ulna; 17 lesser sigmoid cavity. The remaining figures refer to ridges marks, &c., for the attachment of muscles.

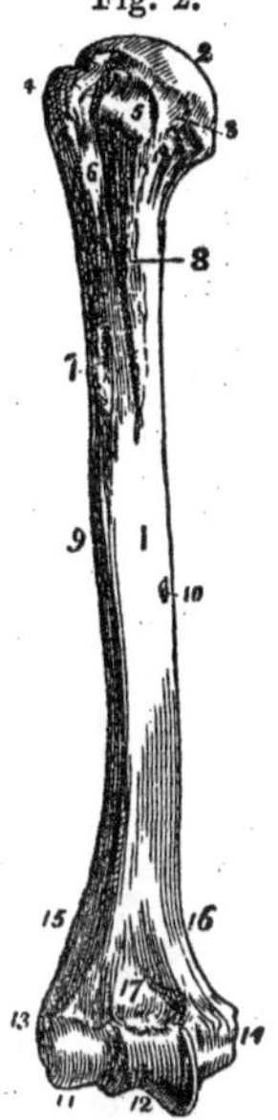
Fig. 2.

The superior extremity or head is hemispherical, and articulates with the glenoid cavity of the scapula; immediately beneath the head, and separated from it by a groove is the neck. Below this groove are two knobs, called tuberosities, one of which is external, and the other internal; they receive the insertion of muscles, and are separated from each other by the bicipital groove, in which the tendon of the long head of the biceps muscle plays.

The inferior extermity is flat and broad, and is covered in front by the brachialis anticus muscle, and behind by the triceps. It presents a hemispherical head for articulating with the radius and an irregular cylindrical surface for the ulna. Just above the articular surface in front is a cavity, called the lesser sigmoid which receives the coronoid process of the ulna when the fore-arm is much flexed. Behind is a corresponding though larger cavity, called the greater sigmoid, for the olecranon process when the fore-arm is extended. Immediately above the articular surface for the radius, and continuous with a ridge three or four inches long is the external condyle, from which and the ridge arise the supinator and extensor muscles. The internal condyle is just above the internal margin of the ulnar articular surface, and is more prominent than the external. It may be felt beneath the skin; from this condyle and the ridge leading from it the flexor muscles of the hand and fore-arm arise.

Bones of the Fore-Arm.

The fore-arm extends from the elbow to the hand, and contains two bones, the ulna, and radius. The ulna is on the same side as the little finger, and is the longer. Both are straight bones.

Fig. 3.

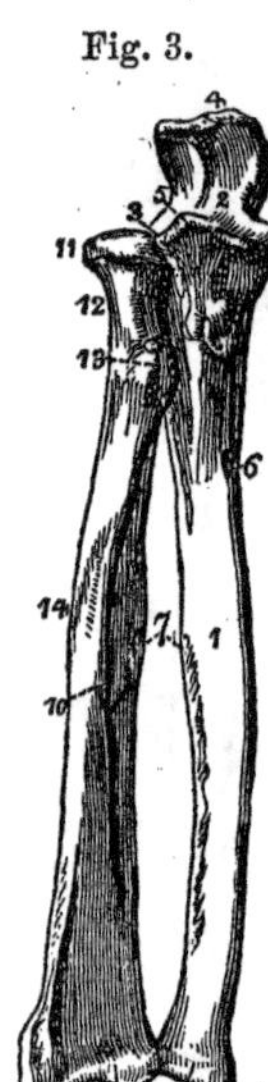

Ulna.—The ulna is somewhat triangular, and its superior or humeral extremity is the larger, and has a hook-like process—the olecranon—behind, to which is attached the triceps extensor cubiti muscle. A little below, and in front of the upper extremity is the coronoid process; the greater sigmoid cavity for articulating with the humerus is between the olecranon and coronoid processes; and on the outside of the coronoid process is the lesser sigmoid cavity for articulating with the radius. The external edge is the sharpest, and to it is attached the interosseous ligament.

Fig. 3. The ulna and radius. 1 body of the ulna; 2 greater sigmoid cavity; 3 lesser sigmoid cavity; 4 olecranon process; 5 coronoid process; 8 lesser extremity; 9 styloid process. 14 Body of the radius; 11 rounded head; 12 neck; 13 tubercle; 15 inferior extremity; 16 styloid process.

Radius.—This bone is placed on the outer side of the ulna, is shorter, and like it extends from the humerus to the wrist. It is slightly curved, and the inferior extremity is the larger. The upper extremity has a cylindrical head, surrounded by a smooth rim or border, a part of which plays in the lesser sigmoid cavity of the ulna while the remainder is in contact with the annular ligament. On the upper surface of the head is a fossa for articulating with the humerus. Below the head is the neck which is about half an inch in length. Immediately beneath the neck is a rough prominence or tubercle for the insertion of the biceps flexor cubiti.

The lower extremity is flattened transversely. It articulates by a concave surface with the scaphoid and lunare bones of the wrist. On the internal face of the extremity is a small articular surface for the ulna. Externally is the styloid process for the attachment of the external lateral ligament. Upon the back of this extremity are several grooves, occupied by the tendons of the muscles which go to the wrist and hand.

The body of the bone is irregularly triangular, and presents three surfaces and three angles, which are principally occupied with the origin and insertion of muscles.

Bones of the Hand.

The hand consists of the carpus or wrist, the metacarpus, and the phalanges, or digiti, and is made up of twenty-seven bones exclusive of the sesamoid bones.

Carpus.—The carpus or wrist is placed next to the fore-arm, and consists of eight bones arranged into two rows, called first and second rows. The bones in the first row are the scaphoid, lunare, cuneiform, and pisiform; those in the second row are the trapezium, trapezoides, magnum, and unciform. They are difficult to distinguish from each other.

The *scaphoides,* placed on the radial side of the wrist, is concave above, and convex below; it articulates above with the radius, below and in front with the magnum, trapezium, and trapezoides, and on the inside with the lunare. It may be known from the others in the row by its greater length.

The *lunare* is at the ulnar side of the latter, and is of a semilunar shape. Above it is convex, and below concave. It articulates with the radius above, with the magnum below, and with the cuneiform on the inside.

The *cuneiform* is placed at the ulnar side of the lunare, and is somewhat wedge-shaped. It articulates above with the lunare, below with the unciform, and on the inside with the pisiform.

The *pisiform,* so named from its resemblance to a pea, is the smallest of the carpal bones. It has but one articular surface, which joins the cuneiform, and may be readily felt at the ulnar side of the arist.

The *trapezium* is placed at the radial side of the second row, and is very irregular in shape; it articulates with the thumb, scaphoid and trapezoides.

The *trapezoides* is at the ulnar side of the trapezium, and the smallest in the second row; it is somewhat pyramidal, with the apex towards the palm, and has four articular faces for the adjoining bones.

The *magnum* is placed at the ulnar side of the last, and is the largest bone in the wrist. On the upper surface it has a rounded head to articulate with the scaphoides and lunare. The body is quadrilateral.

The *unciform* is at the ulnar side of the magnum, and nearly as large; it may be distinguished by its long crooked process.

Metacarpal Bones.

Of these there are five; they are placed between the wrist and the phalanges of the fingers and thumb. Those for the finger are parallel with each other; but the one for the thumb diverges, and is so placed that it can be brought in front of the others during its motions. Each has a rounded head for articulating with its corresponding phalanx, a cylindrical shaft which is smaller than the extremities, and a base which articulates with the carpal bones. The *first*, that for the thumbs, is the shortest and thickest; the *second* for the index finger is the longest; the third, fourth, and fifth gradually diminish in size.

Phalangeal Bones.

Each finger contains three bones, called phalanges; the thumb has but two. The first row, that adjoining the metacarpus, is called the first phalanx; the middle row the second; and the remaining row the third.

The bones of the first phalanx are the largest; they are convex posteriorly, and flattened anteriorly. A superficial cavity exists in the upper extremity of each for articulating with the metacarpal bones.

The bones of the second phalanx are smaller, and are also

convex behind, and flattened in front. The upper extremities of each has two superficial cavities for articulating with the bones of the first phalanx.

The bones of the third phalanx are the smallest, and differ materially from the others. They have but one articular extremity the upper, which has two superficial cavities for the corresponding faces of the second phalanx.

The lower extremity is rounded, flattened, thin, and rough.

The *sesamoid* bones are two in number; they are small and hemispherical, and are placed on the inside of the hand, at the lower extremity of the metacarpal bone of the thumb; they assist the action of the short flexor muscle. Sometimes they are also found at the metacarpal bones of the fingers of robust persons.

Remarks on the Upper Extremities.

The relative proportion of the upper extremities to the lower is much greater at birth than at any after period. This relative size gradually diminishes up to the age of puberty, by which time it has generally disappeared. It is probably owing to the lower extremities, receiving less blood in the fœtal state than the upper.

At birth, also, the extremities of the clavicles, humeri, bones of the fore-arm, carpus, and phalanges—all the long bones—are cartilaginous, and larger proportionably than in the adult. The proportion of animal matter is likewise greater, which renders them less liable to fracture. They may, however, be bent easily, and hence too much force should not be applied to them. Bow-legs are not unfrequently caused by permitting young infants to stand on their feet too long at a time. Distortion of the spine may also be induced by the same cause.

The *clavicle* is susceptible of motion in four directions—upwards, downwards, forwards, and backwards, and also of circumduction, which is a rapid succession of these motions; the articulation at the externum is the centre upon which these movements are performed. It also assists in supporting the shoulder,

9

and by keeping it from falling forwards greatly facilitates the motions of the joint. In females the clavicles are longer in proportion than in males, to accommodate the breasts, and in consequence of this increased length, some of the motions of the shoulder are performed with great awkwardness.

The *scapula,* besides, having all the motions ascribed to the clavicle, is capable of performing a partial rotation; it serves as a movable basis for all the motions of the arm.

The *humerus* is susceptible of motion, upwards, downwards, forwards, and backwards, and also of circumduction and rotation. The motion of circumduction is very extensive in the shoulder joint; it is a regular succession of all the other movements, except rotation, by which the arm describes a cone having for its apex the glenoid cavity.

Rotation is the turning of the bone upon itself; it is not extensive, rarely exceeding a half circle.

In the fore-arm there are two kinds of motion; in one kind the fore-arm is flexed or extended on the arm, and the ulna is the principal agent. In the other, the radius rotates upon the ulna, the latter being almost stationary. When the hand, following the forward rotation of the radius, has its palm directed downwards, it is said to be in a state of supination. This is the most common and easiest position of the fore-arm. In the backward rotation of the radius, in which the palm is upwards, the hand is in a state of supination; in this position the radius is parallel with the ulna; in pronation the middle part of the radius crosses that of the ulna.

The motions of the hand on the fore-arm are, flexion, extension, lateral inclination, or abduction, and adduction, and circumduction; those which take place between the first and second rows of the carpus are chiefly flexion and extension. The metacarpal bone of the thumb moves freely on the trapezium; its motions are, flexion, extension, abduction, adduction, and circumduction. The latter is the result of the others, and resembles that of the shoulder joint. The motions of the other metacarpal bones are much more limited, being confined mostly to a moderate degree of flexion and extension. The first phalanges admit of flexion, extension, lateral motion, and circumduction; the second and third are restricted to the two first. It will thus be perceived from the nice mechanism of the upper extremities, and the number of pieces which enter into their structure, what a variety of movements they are capable of performing.

Lower Extremities.

The bones of the lower extremity are, the femur, os femoris, or thigh bone, the tibia, fibula, patella; and the tarsal, metatarsal, and phalangeal bones belonging to the foot—in all sixty-four.

Thigh Bone, or Femur.

The thigh contains but one bone, which is the longest in the body, reaching from the pelvis to the knee; it consists of a superior, and an inferior extremity, and a body.

Fig. 4.

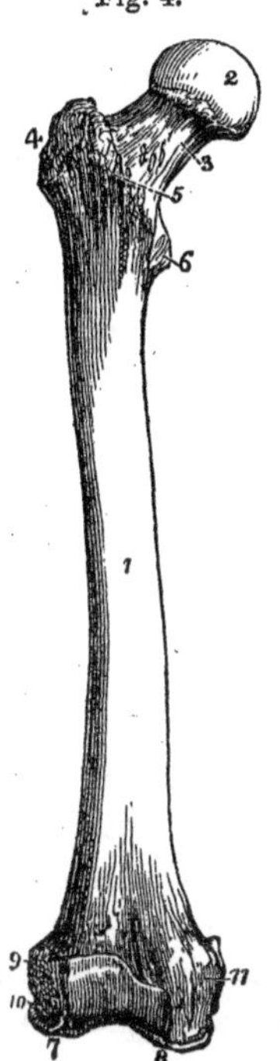

Fig. 4. Thigh bone, or Femur. 1 Body, or shaft; 2 head; 3 neck; 4 trochanter major; 6 trochanter minor; 7 internal condyle; 8 external condyle.

The superior extremity has a spherical head, with a depression upon it for the ligamentum teres. Between the head and body at an angle of about thirty degrees with the latter is the neck, which is about two inches in length. At the base of the neck on either side are two prominences, called the trochanters major and minor; the former, which is situated externally, is much the larger, and receives the insertion of several muscles coming from the pelvis; the latter is placed internally, and has the psoas magnus and iliacus internus muscles inserted into it. Between the two trochanters behind is a ridge, into which is inserted the quadratus femoris. A similar ridge in front marks the attachment of the capsular ligament.

The inferior extremity is much larger than the superior, and is divided by a fossa in front and a notch behind into two parts, called the

Fig. 5.

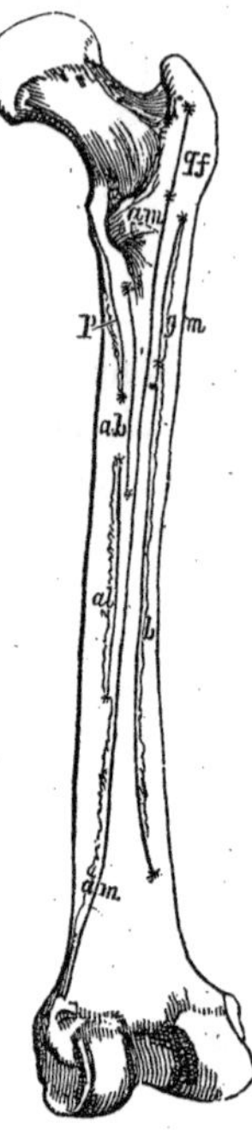

external and the internal condyle. Of these the internal from being placed the most obliquely, appears to be the larger. These condyles have articular faces for the tibia below, and the patella in front.

The body of the bone is nearly round, and is slightly curved anteriorly; its front surface is covered by the cruræus muscle.

Fig. 5. Posterior view of femur, showing the ridge, called the linea aspera, and the origin and insertion of the muscles along its two lips.

p insertion of the pectineus; a b insertion of the adductor brevis; g m insertion of the gluteus maximus; a m insertion of the adductor magnus; g m origin of the vastus externus; b origin of the short head of the biceps flexor cruris.

Posteriorly there is an elevated ridge, the linea aspera, which is considerably elevated, and serves for the origin and insertion of muscles. Near the middle of this ridge is a large foramen for the nutritious artery.

Bones of the Leg.

There are two bones to the leg, the tibia and the fibula, of which the former is placed internally, and the latter externally; they extend from the knee to the foot.

Tibia.—This bone is much longer and larger than the fibula; next to the femur it is the largest bone in the body; its lower half is commonly called the shin-bone. The upper extremity, or head, is large, and contains two superficial cavities, divided by a ridge, called the spinous process, for articulating with the condyles of the femur. At the base of the spinous process in front and behind the crucial ligaments are attached; at its summit is a depression for the attachment of the posterior extremity of the external semilunar cartilage. Upon each side of the head is an enlargement, called the internal and external condyles, or tuberosities; on the back part of latter, looking downwards, is a small articular mark for the fibula. In front and just below the head is a tubercle for the attachment of the ligament of the patella.

The lower extremity of the tibia is much smaller than the upper, and somewhat quadrilateral; it articulates by a transverse cylindrical concavity with the astragalus; externally there is a triangular groove for articulating with the fibula; internally a large process, called the internal malleolus; posteriorly a slight groove for the tendon of the flexor longus pollicis muscle. The extensor tendons pass over the anterior surface.

The body is triangular, and consequently presents three edges and three surfaces. The anterior edge, called the spine, or crest, is sharp, superficial, and slightly curved; the external has attached to it one edge of the interosseous ligament; the internal edge is rounded, and in it are inserted several muscles. The internal surface is only covered by the skin; the external surface is covered by the muscles of the leg; the posterior surface gives origin to the tibialis anticus and flexor communis muscles.

Fig. 6.

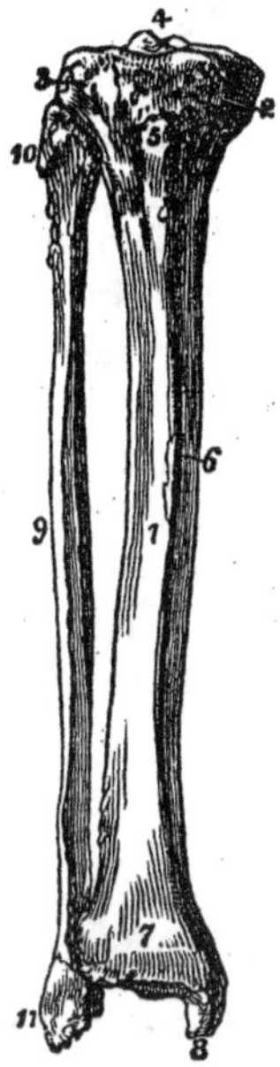

Fig. 6. Anterior view of the tibia and fibula. 1 External face of the body of tibia; 2 internal condyle; 3 external condyle; 4 spinous process; 5 tubercle; 6 anterior edge, or spine; 7 lower extremity; 8 internal malleolus; 9 fibula; 10 upper extremity; 11 lower extremity, or external malleolus.

Fig. 7.

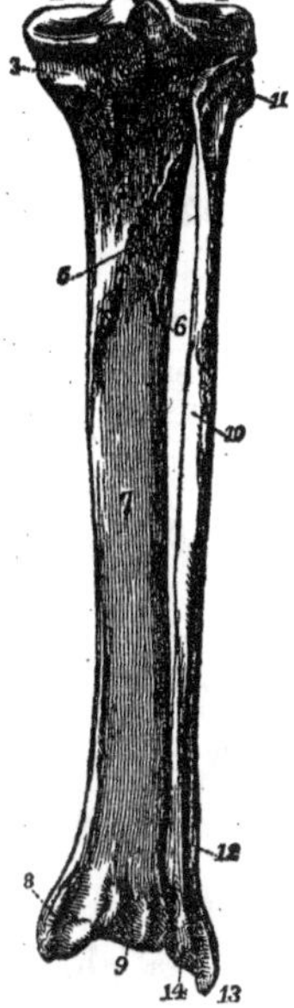

Fig. 7. Posterior view of the tibia and fibula 1 External condyle; 2 internal condyle; 3 fossa for the insertion of the semi-membranosus muscle; 4 fossa for the attachment of the posterior crucial ligament; 5 oblique ridge for the origin of the soleus muscle; 6 external edge; 7 posterior face of body; 8 internal malleolus; 9 groove for the tendon of the flexor longus pollicis; 10 fibula; 11 superior extremity; 13, 14 inferior extremity, or external malleolus.

Fibula.—The fibula is placed on the outside of the leg, with its upper part somewhat posteriorly. It is a long, thin bone, extending from the head of the tibia to the foot; it is rather shorter than the tibia. The upper extremity or head is enlarged, and has a concave surface on its upper part for articulating with the external condyle of the tibia; behind is a styloid process for the insertion of the head of the biceps. The lower extremity is also enlarged, and is called the external malleolus; its internal surface is smooth, triangular, and slightly convex for articulating with the astragalus; its external surface is also triangular and superficial. The pointed extremity of the external malleolus is sometimes called the coronoid process.

The body is triangular, presenting a twisted appearance, and has three faces; the external face gives origin to the peroneus longus and brevis muscles in its upper two-thirds; the internal face is towards the tibia, and is divided longitudinally by a ridge to which is attached the interosseous ligament; the spaces in front and behind this ridge give origin to muscles; the posterior face gives origin to the soleus and flexor longus pollicis muscles. The anterior angle is sharp and elevated in the middle. There is also a slight bend, towards the tibia, in this bone.

Patella.—This is a flat ovoid bone, situated at the fore part of the knee joint; it is commonly called the knee-pan. Its anterior face is convex and rough, and covered by the integuments; its posterior face is smooth, and unequally divided by a longitudinal ridge. The external part is the larger, and articulates with the trochlea in front of the external condyle of the femur; the internal part is the smaller, and articulates with the trochlea of the internal condyle. The superior margin is thick, and into it is inserted the tendon of the rectus femoris; the inferior margin is thinner and pointed, and has the tendon of the patella attached to it.

Bones of the Foot.

The foot consist of the tarsus, metatarsus, and phalanges. In the *tarsus*, which forms the posterior half of the foot, are seven bones, the os calcis, astragalus, naviculare, cuboides, and the external, middle, and internal cuneiform.

Os calcis.—This bone forms the heel, and is much the largest of the bones of the foot. In shape it is irregular; its longest diameter is lengthwise of the foot, and it is also thicker vertically than transversely. The superior surface has two articular faces at its front part for the astragalus. Between these cavities is a deep groove, the posterior part of which is occupied by the interosseous ligament. The internal surface is very concave, and is called the sinuosity; over it pass the tendons, vessels, and nerves for the sole of the foot; the external surface is nearly flat, and is marked by the passage of tendons of the peroneus longus and brevis; the under surface is slightly concave, and has two tuberosities behind, of which the internal is the larger; they give origin to the muscles of the foot. This surface has also a tuberosity in front.

The anterior extremity forms the greater apophysis, and articulates with the cuboides; the posterior extremity is convex, and rough to receive the insertion of the tendon Achillis.

Astragalus.—This is next in size to the latter, and is placed between it and the bones of the leg. It consists of a body and head. Above, the body articulates with the tibia, and on either side with the malleoli. The head is placed anteriorly and articulates with the scaphoides.

Scaphoid.—This is an oval bone situated at the inner side of the tarsus, between the astragalus and the cuneiform bones; behind is a deep concavity for the astragalus, and in front three triangular articular surfaces for the three cuneiform bones; at the inner side is a large tuberosity for the insertion of the tendon of the tibialis posticus.

Cuboid.—Is on the outer side of the tarsus, between the os calcis and the metatarsal bones, and as its name indicates, is somewhat cuboidal in shape. The internal face is flat, and has an articular mark for the cuneiform internum; the posterior face is triangular, and articulates with the os calcis; the anterior face articulates with the last two metatarsal bones.

Internal cuneiform.—This is the largest of the three cuneiform bones, and is placed between the scaphoid and the first metatarsal bone. Anteriorly it joins the first metatarsal bone; posteriorly the scaphoid; internally it is marked by the tendon of the tibialis posticus; externally it joins the second metatarsal bone, and middle cuneiform. As may be infered from the name, in its general shape it resembles a wedge.

Middle cuneiform.—Is the smallest of the tarsal bones, and is also shaped like a wedge with the base above; it is placed upon the scaphoid, at the outside of the internal cuneiform. It articulates with these two bones and likewise with external cuneiform.

External cuneiform.—This is also wedge-shaped, and placed upon the scaphoid, between the second cuneiform and the cuboides. It articulates anteriorly with the third metatarsal bone; posteriorly with the scaphoid; internally with the second cuneiform, and the second metatarsal bone; externally with the cuboides.

Bones of the Metatarsus.

In the metatarsus are five parallel long bones; they extend from the tarsus behind to the toes in front, and are called numerically, beginning on the inner side of the foot.

The *first* is shorter and thicker than the others; its base articulates with the internal cuneiform, and its head, which is spherical, with the first phalanx of the great toe. Below its head are the sesamoid bones.

The *second* is the longest; its base articulates at the extremity with the middle cuneiform, on the inside with the internal cuneiform, and on the outside with the external cuneiform and second metatarsal.

The *third* has a triangular base which articulates with the third cuneiform, and also laterally with the second and fourth metatarsal bones.

The *fourth* articulates at its base with the cuboid, and on either side with the third and fifth metatarsal bones.

The *fifth* is the shortest, and is readily distinguished by a large tubercle, which projects outward from its base beyond the external margin of the cuboides, into which the tendons of the peroneus tertius and brevis are inserted. Its base articulates with the cuboid and fourth metatarsal bones; its anterior extremity is more rounded than that of the others.

Bones of the Toes.

There are five toes on each foot, which are named numerically, beginning at the great toe. Each toe is formed of three bones, called phalanges, except the great toe, which like the thumb, has but two.

The first row or phalanx bear a general resemblance to the first row of the fingers, but are smaller; the bases articulate by a deep cavity with the metatarsal bones; the anterior extremities have two small condyles for the second row.

The second row or phalanx, also resemble, the second row of the fingers, though they are much shorter, scarcely having any bodies; the bases have two cavities for articulating with the first phalanx, and the anterior extremities two convexities for the third phalanx.

The third row articulates with the second, and are very small; the fourth and fifth are generally but imperfectly developed.

The phalanges of the great toe are much larger than the others. Two sesamoid bones are connected with the tendon of the flexor brevis pollicis of each foot; they have articular faces which join the head of the metatarsal bone of the great toe.

Remarks on the Lower Extremities.

As before remarked the lower extremities are proportionably smaller than the upper at birth, owing to their receiving up to that time a smaller quantity of blood.

Like those of the upper extremities, the bones of the lower are but partly ossified at birth; the ends of all the long bones are cartilaginous, as are also the patella and tarsal bones with the exception of parts of the os calcis and astragalus. The metatarsal and phalangeal bones are not so much developed as the corresponding bones of the hand.

At birth also the upper end of the femur is more at a right angle to the body than it is in the adult; the body of the bone is but slightly curved, and the neck short; all of which tend to make standing and walking difficult in young children.

The bones of the lower extremities all become ossified about the fifteenth year, and they have then nearly the same form as in the adult. The epiphyses are the only exceptions to this, which where they join the body of the bone are often cartilaginous, and until the twentieth or twenty-fifth year may be separated by boiling.

The motions of the thigh are the same as those of the arm, viz. flexion, extension, abduction, adduction, circumduction, and rotation, though they are much less extensive in consequence of the acetabulum, the basis on which they are performed, being fixed.

Flexion, by which the thigh is carried forward, is performed with great facility and ease; extension, the reverse of this, is also performed with much freedom, though it is less extensive than the former; abduction is that motion by which the thighs are separated, and adduction the act of bringing them together again or crossing them; the muscles concerned in the latter are called adductors. Both of these motions are performed to a considerable extent. Circumduction—the regular succession in a circle of the other motions mentioned—is less extensive than in the arm. Rotation or the turning of the femur upon itself, is effected with much facility, especially rotation outwards, or backwards.

The chief motions of the leg on the thigh are flexion and extension; rotation is also performed in a slight degree. Those of the foot upon the leg are flexion, extension, and some lateral motion.

Of the Articulations.

The connection of one bone with another is called an articulation, or joint. Some articulations are movable, and others immovable; in the structure of the former cartilage, ligaments, and synovial membranes are necessary.

Cartilage is a smooth, white, flexible, and elastic substance, and ranks next to bone in hardness. Chemically, it is composed of gelatine and water with a small portion of phosphate of lime. It resists the attacks of disease almost as firmly as bone; it is not highly organized, being without red blood-vessels, nerves, and lymphatics, and unless diseased has no sensibility. In old age it is disposed to ossify. It is invested by a fibrous membrane, called *perichondrium*, which corresponds with the periosteum.

That portion of cartilage, called *articular*, which covers the extremities of bone, is thicker in the middle when it covers convex surfaces, and thicker on the edges when it lines cavities.

Cartilages, which are movable in a joint, are called *inter-articular*.

Fibro-cartilage, which consists partly of cartilage, and partly of ligament, and is much stronger than the former, is found in the external ear, between the vertebra, in the knee-joint, &c.

Ligaments are composed of fibrous tissue, and are inelastic. Those connecting the joints pass from one extremity of the bone to the other, mostly exterior to the synovial membrane. They are of two kinds, white and yellow; the tendons and most of the ligaments are examples of the former, and the ligamentum nuchæ of the latter. When ligaments are cord-like, they are called funicular, and when open at the ends like a bag, capsular.

Synovial membranes are thin, closed serous sacks, which line the movable joints. They secrete a viscous fluid, resembling the white of an egg, called *synovia*, which lubricates the joints and prevents friction.

The different kind of articulations have been divided into Synarthrosis, Amphiarthrosis, and Diarthrosis.

Of the first, which implies immobility, there are several species—that which unites the bones of the skull is termed *Sutura;* that which unites the bones of the upper jaw *Harmonia;* that by which the vomer joins the azygos, *Schindylesis;* and that joining the teeth with the alveoli, *Gomphosis.*

In the second, which implies partial motion, the bodies of the vertebræ and symphyses are included.

In the third, diarthrosis, three species are included; *Arthrodia,* that uniting the tarsal and carpal bones; *Ginglymus,* or hinge-like, as the elbow and wrist; and *Enarthrodia,* or ball and socket joint, as the hip and shoulder.

In the *articulation* of the *lower jaw* there are: a capsular ligament, an internal and an external lateral ligament, a stylo-maxillary ligament extending from the styloid process of the temporal bone to the angle of the jaw; an inter-articular cartilage—a thin oval plate dividing the joint into two cavities—with two synovial membranes, one on either side of the inter-articular cartilage; or when the cartilage is imperfect but one synovial membrane.

In the *articulation* of the *vertebræ* there are: an anterior vertebral ligament, which is placed in front of the bodies of the vertebræ, and extends from the second cervical to the first sacral vertebra, gradually increasing in breadth as it descends; a posterior vertebral ligament, which is placed at the posterior part of the bodies of the vertebra within the spinal canal, and extends from the occiput to the coccyx; it is narrow and thick in the thorax, wider below, and adheres more closely to the inter-vertebral substance than to the bodies of the vertebræ, which gives it a serrated appearance. Between the bodies of the vertebræ, and adhering closely to their substances, are twenty-three fibro-cartilaginous disks; these are arranged in concentric lamina, compressible, and gradually increase in thickness from above downwards. Owing to this compressibility, the trunk becomes shortened after the erect position has been maintained for several hours, but is restored to its original length again by an interval of rest in the horizontal posture.

The *oblique processes* are united by a capsular ligament.

The *spinous processes* are joined together by the inter-spinal ligaments, which fill up the spaces between them. They are, however, wanting in the neck, and their place is supplied by the ligamentum nuchæ, which extends from the seventh cervical vertebræ to the posterior occipital protuberance, dividing the muscles of the neck.

The *yellow ligaments*, which are elastic, are placed between the bony bridges of the vertebræ; there are twenty-three pairs of them.

Between the occiput and atlas are: an anterior ligament, which extends from the back parts of the occipital foramen to the front of the atlas; a posterior ligament, which extends from the back part of the occipital foramen to the corresponding edge of the atlas; and a capsular ligament, surrounding the superior oblique process of the atlas, and the condoloid process of the occiput.

Between the atlas and dentata are: a transverse ligament, which passes across the atlas transversely from one tubercle to the other; two moderator ligaments, extending from the sides of the processus dentatus to the inner side of each occipital condyle; a middle or straight ligament, passing from the processus dentatus to the anterior edge of the occipital foramen; a loose capsular ligament surrounding the oblique processes; and some ligamentous bands, called Lacerti ligamentosi, reaching from the back part of the body of the dentata to the occiput.

In the *articulation* of the bones of the *pelvis*, there are, uniting the sacrum and ilium, an anterior sacro-iliac ligament, which consists of short fibres passing from one bone to the other; a posterior sacro-spinous ligament extending from the spinous processes of the ilium to the third and fourth transverse processes of the sacrum; and an anterior and posterior sacro-sciatic ligament; the posterior, which is much the larger, arises from the posterior inferior spinous process of the ilium, the margin of the sacrum, and the first bone of the coccyx, and is inserted into the inner margin of the tuberosity of the ischium; the anterior, which is in front of the posterior, extends from the side of the sacrum to the spine of the ischium. The articulating surfaces of both the sacrum and ilium are covered with cartilage.

The obturator ligament passes over the thyroid foramen

closing it up, with the exception of a small opening at its upper part for the passage of the obturator vessels and nerves.

In addition to the foregoing there is an ilio-lumbar ligament, a lumbo-sacral ligament, an anterior and posterior coccygeal ligament, and a sub-pubic ligament, the names of which sufficiently indicate their attachments.

The articulation of the pubes—symphysis pubis—consists chiefly of a fibro-cartilaginous substance, like that of the vertebræ.

The *Ribs* in their articulation posteriorly with the vertebræ have an anterior or radiated ligament, extending from the head of the rib to the two contiguous vertebræ and the intervening cartilage; a capsular ligament enclosing the head of the rib; an inter-articular ligament, extending from the ridge on the head of the rib to the inter-vertebral substance; a capsular ligament connecting the transverse processes to the tubercle of the rib; an internal, middle, and external costo-transverse ligament, extending from the transverse process to the adjacent rib.

In the articulation of the ribs anteriorly with the sternum there is: an anterior radiated ligament, which passes from the cartilages of the true ribs to the sternum; a posterior radiated ligament, having the same connections; and a costo-zyphoid ligament, which extends from the cartilages of the sixth and seventh ribs to the zyphoid or ensiform cartilage.

In the articulation of the *shoulder* there are: uniting the clavicle and sternum, a capsular ligament, within which, and dividing the joint into two cavities, is a wedge-shaped inter-articular cartilage; an inter-clavicular ligament, connecting the two clavicles; and a rhomboid ligament, extending from inferior edge of the clavicle to the cartilage of the first rib.

Uniting the scapula and clavicle are: a capsular ligament, connecting the acromion process of the scapula, and the outer end of the clavicle; a coraco-clavicular ligament, consisting of two parts, one called conoid, the other trapezoid, which extends from the coracoid process to the external extremity of the clavicle; a bifid ligament, which goes from the coracoid process to the clavicle and the cartilage of the first rib; a coraco-acromial ligament connecting the coracoid and acromion processes; and a coracoid ligament which passes across the coracoid notch.

Uniting the scapula and humerus, are, a loose capsular ligament which invests the glenoid cavity and the neck of the humerus; a coraco-humeral ligament, a part of the capsular, which extends to the coracoid process; and a glenoid ligament which consists of a ring of fibro-cartilage attached to the edge of the glenoid cavity, to increase its depth.

In the articulation of the *elbow joint* there are, a capsular ligament, which surrounds the head of the humerus, radius and ulna; an internal lateral ligament, passing from the internal condyle to the coronoid and olecranon processes of the ulna; an external lateral ligament, extending from the external condyle to the lateral ligament; an angular or coronary ligament, which surrounds three-fourths of the head of the radius, and extends to either side of the lesser sigmoid cavity; and an interosseous ligament, which fills up the space between the radius and ulna throughout nearly their whole length.

The articulation of the *wrist,* which is formed by the greater sigmoid cavity of the radius, and the scaphoid, semilunar and cuneiform bones, has a capsular ligament, an anterior and a posterior ligament, and an internal and external lateral ligament. The individual bones of the carpus are connected by palmar and dorsal ligaments.

The *bases* of the metacarpal bones are also attached to the second row by palmar and dorsal ligaments. The pisiform bone and the thumb have distinct capsular ligaments.

The fingers are attached to the metacarpal bones by an internal and external lateral ligament; an anterior or palmar ligament, and a posterior ligament furnished by the extensor tendon.

The phalangeal bones are joined together by a similar arrangement of ligaments.

In the articulation of the *hip joint,* formed by the head of the thigh bone and the acetabulum, the following ligaments enter. The cotyloid ligament, which is a thick prismatic ring surrounding the margin of the acetabulum and increasing its depth. The ligamentum teres—round ligament—which is attached to a pit on the head of the femur, whence it passes in two fasciculi to be inserted into the notch of the acetabulum; the cotyloid ligament and the capsular ligament, which encircles the acetabulum and the neck of the femur, and is the strongest ligament in the body.

In the articulation of the *knee joint* there are, an anterior ligament, called also the ligament of the patella, which is a continuation of the quadriceps muscle, is firmly attached to the patella, and inserted into the tubercle of the tibia; a posterior ligament, also called ligament of Winslow, which extends obliquely from the external condyle to the back part of the internal tuberosity of the tibia; two semilunar cartilages, which are prismatic rings attached to the margins of the tibia, to deepen its articular surface; two crucial ligaments, which cross each other, the one extending from the front of the spine of the tibia to the posterior part of the inner face of the external condyle, the other passing from behind the spine of the tibia to the anterior part of the external face of the internal condyle of the femur; and an internal and external lateral ligament on either side of the joint.

The *tibia* and *fibula* are united; at the superior extremity, by an anterior and posterior ligament, which pass obliquely between the heads of the two bones before and behind; along their length by an interosseous ligament, which fills up the space between the bones, except a small opening at the upper extremity to transmit the anterior tibial artery; and at the lower extremity, by an anterior and posterior ligament similar in arrangement to those connecting the heads of the bones.

The *ankle joint* is united by an internal and an external lateral ligament; the internal, also called deltoid is triangular in shape, and has its apex attached to the internal malleolus, and its base to the os calcis, astragalus, and calcaneo-scaphoid ligament; the external consists of three fasciculi which arise from the external malleolus, and are inserted into the astragalus and os calcis.

Articulation of the tarsus.—The astragalus is united to the os calcis by a thick, strong, interosseous ligament, and by a posterior ligament; the os calcis to the scaphoid by a superior and an inferior calcaneo-scaphoid ligament; the os calcis and cuboid by a superior and inferior calcaneo-cuboid ligament; and the astragalus and scaphoid by a semicircular ligament which passes from the neck of the astragalus to the edge of the concavity of the scaphoid. The three cuneiform bones are joined to the scaphoid, and to each other by dorsal, plantar and interosseous ligaments.

Articulation of the metatarsus.—The first of these bones has a strong capsular ligament, attaching its base to the internal cuneiform; the second and third have the base attached to the middle and external cuneiform by dorsal and plantar ligaments; and the bases of the fourth and fifth are joined to the cuboid by dorsal and plantar ligaments. The heads of the metatarsal bones are united to the phalanges by two lateral and a plantar ligament, and superiorly by an expansion of the extensor tendon; and to each other by a strong transverse ligament.

In the articulation of the phalanges there is an arrangement of ligaments corresponding to that of the hand.

Of the Skin.

The skin covers the entire body, and besides affording protection to the various parts, serves also as an organ of touch and secretion.

It is continuous with the mucous membrane, and at the orifices of the canals leading into the body, as the mouth, nose, &c., is readily converted into it. Its color and thickness vary in different races and individuals, and in different parts of the same individual. Climate has a great influence over it, the tendency of parts exposed to tropical heat and light being to turn dark.

The skin is covered by a great number of wrinkles; the largest, those which appear on the forehead, face, &c., are caused by the contractions of muscles, and the flexion of the joints; others of a finer description are produced by the contractile nature of the skin.

Numerous hairs, perspiratory ducts, and pits showing the orifices of sebaceous glands, and follicles, are contained in the skin; the perspiratory ducts or pores are not visible to the naked eye. The facility with which the skin slides backwards and forwards on most parts of the body is owing to the looseness of the cellular tissue beneath it.

The skin consists of two layers, the cutis vera or true skin, also called chorion, and the epidermis or cuticle. Formerly a third layer, the rete mucosum, was admitted; but more recent observations have demonstrated this to be merely a layer of the cuticle.

Of the two layers the *true skin* is the thicker and deeper; it is closely adherent to the cellular tissue, and perfectly white and semi-transparent in all persons. The external surface is covered with fine conical projections, called villi or *papillæ tactus;* the papillæ are most numerous and distinct in those parts where there is much motion. On the hands and feet they are arranged in double rows, which occasion the semi-circular and spiral wrinkles of the cuticle on these parts. Each papillæ consists of an artery, vein, and nerve, and the sensibility of a part is in proportion to their number.

The texture of the true skin is fibrous; the irregular interlacing of fibres producing a mass of net-work or areolæ, through the meshes of which the hairs, nerves, blood-vessels, &c., are transmitted. Its composition is of condensed cellular tissue, the yellow fibrous element predominating where great elasticity is required, as in the arm pit; and the white element where resistance is wanted, as in the sole of the foot. It unites readily with tannin, and forms leather. Boiling converts it into gelatine.

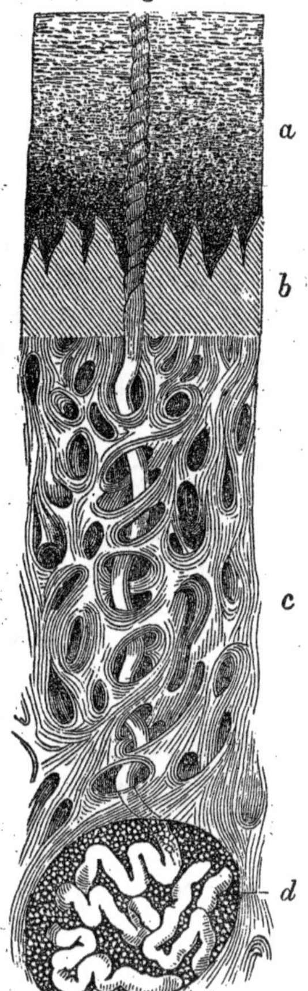

Fig. 8. Magnified representation of the skin. a, Cuticle or epidermis; b, papillæ tactis or villi; c, cutis vera or true skin.

The *cuticle* or *epidermis* varies in thickness according to the amount of pressure to which it is subject; in the palms of the hand, and the soles of the feet in individuals, who perform manual labor, it becomes very thick. It is not organized, neither vessels nor nerves being traceable into it, but is composed of particles arranged in laminæ, or layers; the deepest being granular, the next somewhat flattened, and those upon the surface mere horny scales which are constantly being shed, and replaced again by others. The coloring matter is contained in the lowermost layer, and is very abundant in negroes, moles, freckles, &c. The facility with which the cuticle may be divided into different layers, owing to its laminated arrangement, formerly caused the innermost layer to be considered distinct, and to receive a separate name, that of rete mucosum.

Of the Sebaceous Glands.

These consist of a blind, pouch-like duct, which is lined by an epithelium containing granules of sebaceous matter in its particles, and has an orifice opening into hair follicles, or upon the general surface. They are most abundant on the scalp and face, particularly about the nose. There are none on the palms of the hands and the soles of the feet.

The secretion is of an oily nature, and serves to lubricate the hair and skin; it is this which gives to linen that has been worn a long time a greasy appearance, causes water when applied to the surface of the body to collect in drops, and gives rise to the strong disagreeable smell in negroes, and persons who do not pay proper attention to cleanliness. Parasitic animalculæ are frequently found in the ducts of these glands.

Of the Nails.

The nails take the place of the cuticle on the ends of the fingers and toes, and like the latter are readily separable from the true skin by maceration. They correspond with the hoofs and talons of the lower orders of animals.

They consist of a root, body, and free extremity. The root is that part which is concealed; it has a thin irregular edge, is white, thin, soft, and about one-fifth of the whole length of the nail. The body is that portion between the root and free extremity; its under surface adheres closely to the skin, which produces it, and is hence called the matrix; it is also soft and marked by longitudinal grooves.

The white part of the nail near the root is called the crescent or lunula, and is caused by the absence of vascularity. At its root the nail is received into a groove formed by the cutis vera.

Of the Hairs.

Hairs are found on almost every part of the skin except the palms of the hands and soles of the feet. They vary in size and color in different races, sexes, and individuals, and in different parts of the body. Those on the head attain the greatest length, and grow more closely together than any others—in females they grow longer, and are more abundant than in males. Those of the face (the beard), when allowed to grow, are next in length, and are thicker than the others, and more disposed to curl. Generally the eyes and hair correspond in color, and the darker the hair the coarser. It has been computed that a fouth of an inch square has upon it 147 black hairs, or 162 hazel, or 152 white. In some individuals the hairs are so much developed as almost to conceal the skin.

Each hair consists of a bulb and a shaft or stalk; the bulb is the extremity contained within a follicle of the skin, and the shaft is the part that projects beyond the surface. The *human* hair is solid and not globular as is generally supposed. It is continuous with the cuticle lining the follicle, and is arranged in lamina or scales, which overlap each other like the shingles of a house; the middle portion has a looser and more porous structure than the external, and has received the name of medulla, while the external is termed the cortex. The hair is nourished from the bottom of the follicle, as the nail is from the matrix, or the cuticle from the true skin.

The hairs, when large, are used as organs of touch, and in some animals, the whiskers of the cat, for instance, possess some degree of sensibility, which is owing to the projection of a conoidal papilla of the follicle furnished with nerves, into the bottom of the bulb. As a general rule, however, they are void of sensibility. The hairs receive no blood-vessels; their moisture is chiefly owing to the secretion of the sebaceous follicles passing through them by capillary attraction. In those instances in which the hair has suddenly become white from mental emotion, it has been by some, supposed to be owing to the secretion of a fluid acid, which penetrates the tissue of the hair in this way, and destroys its color.

The erection of the hair in some animals is caused by the contraction of the subcutaneous muscle; and a similar effect in man, produced by great fright, is owing to the contraction of the occipito frontalis muscle. Moisture lengthens the hairs and dryness shortens them.

When the hair falls off, it is occasioned by the drying up of the fluids of the follicle or its death; in the baldness of old persons, the latter generally takes place, and consequently all efforts to restore it will be fruitless. But when the hairs fall off from sickness, the follicles remain, and a new crop is soon produced. When the hair becomes white from age, the change of color begins at the free extremity; but if the color is restored again, the change begins at the root. In the disease known as Plica Polonica, which consists in a matting together of the hair by a glutinous substance from the cutaneous glands, if the hairs are cut close to the skin they bleed; this is owing to the elongation of the vascular papilla at their roots by disease.

Of the Teeth.

The permanent teeth are thirty-two in number, sixteen being placed in the alveolar processes of each jaw. They are the hardest part of the human body. The part of a tooth above the gum is called its crown; the narrow portion surrounded by the gum its neck; and the portion within the alveolus its fang, or root. There are four classes of teeth, as follows: two *incisors,* or cutting teeth, on each side of each jaw next the middle line; one *cuspid,* or pointed place next to these outwardly; two *bi-cuspid,* or double pointed; and three *molars,* or grinding teeth.

Fig. 9. Permanent teeth.

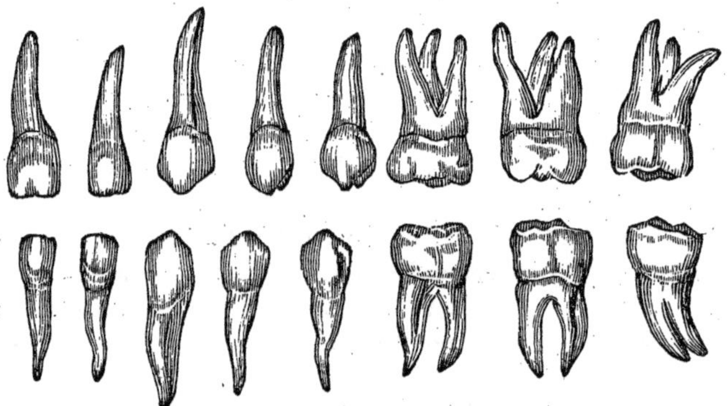

The *incisors* have a bevelled edge which in early life is serrated; their roots are single and conoidal; those of the upper jaw are the largest.

The *cuspid* (pointed), or canine, which are placed between the incisors and bi-cuspides, have a conoidal body and longer roots than any of the others. Those of the upper jaw are commonly known as eye-teeth, and those of the lower as stomach-teeth.

The *bi-cuspides* are next in size to the molars; their bodies have two conical tubercles, or grinding points, the external of which is the larger. The root is flat and deeply grooved on either side. The two posterior superior have each two roots; the others generally but one.

The *molars*, of which there are six in each jaw, have large quadrilateral bodies, surmounted by four or five grinding points; their roots are shorter than those of the bi-cuspides, and are divided into two, three, four, or even five branches. The third, called the wisdom-tooth, is generally smaller than the other two.

In the centre of each tooth is a cavity, which is filled by a *pulp*, chiefly composed of an artery, vein, and nerve, which enter through the small orifice at the extremity of the root. This pulp is soft, of a gray color, and extremly sensitive.

Three textures enter into the *structure* of each tooth—the ivory, or bony portion, the enamel, and the cementum. The first, which forms the principal part of the tooth, is arranged in radiating fibres; it contains neither blood-vessels, nor nerves, nor are its particles absorbed as those of bone. Chemically, it consists of phosphate of lime, gelatine, and water. The *enamel* is the hardest portion, it covers the body of the tooth, is white, brittle, and semi-transparent, thicker on the grinding surface, and terminates by a thin edge at the neck. It is also arranged in radiating fibres, and has neither blood-vessels, nor nerves. The *cementum* is a thin coating extended over the root of the tooth, which in structure resembles bone.

The periods at which the permanent teeth make their appearance are as follows: the four first molars, and the two inferior incisors from the 6th to the 8th year; the two superior central incisors from the 7th to the 9th year; the four lateral incisors from the 8th to the 10th year; the four first bi-cuspides from the 9th to the 11th year; the four cuspidati from the 12th to the 13th year; the four second bi-cuspides from the 11th to the 13th year; the four second molars from the 12th to the 14th year; and finally the four last molars from the 18th to the 30th year, or even later in some instances.

Fig. 10.

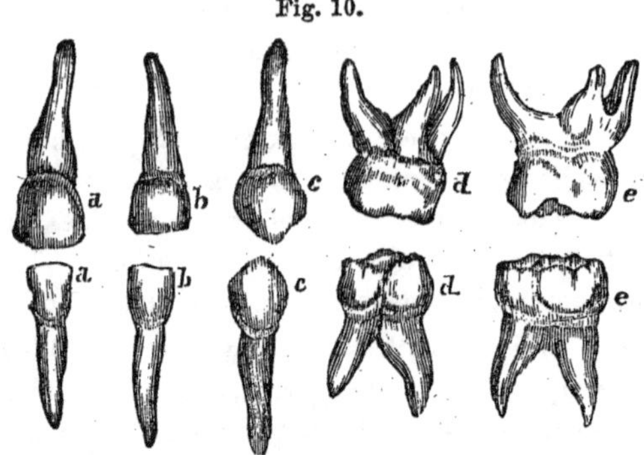

Fig. 10. The deciduous, or milk-teeth. a a central incisors; b b lateral incisors; c c cuspidati, or eye- and stomach-teeth; d d first molars; e e second molars.

The rudiments of all the teeth, both of the first and second set, fifty-two in number, are contained in the gum of the infant. The first set, called the *deciduous*, or milk-teeth, are twenty in number, viz. four incisors, two cuspidati, and four molars in each jaw. As with the temporary teeth there is a good deal of irregularity in the time of their appearance; usually the four central incisors appear from the 5th to the 10th month; the four lateral incisors from the 8th to the 13th month; the four first molars from the 10th to the 14th month; the four cuspidati from the 12th to the 18th month; and the four second molars from the 24th to the 30th month. The teeth of the lower jaw generally make their appearance a short time before those of the upper. About the seventh year, the time at which the first of the permanent teeth make their appearance, the first set begin to loosen and fall out, generally, in the order in which they came.

Of the Muscles.

The muscles are the organs by which the various motions of the body are performed; they are constituted of the substance commonly called flesh, and possess the power of shortening themselves, called *contractility*, and which produces motion.

Each muscle is composed of a number of *fasciculi* (bundles of fibres) of various size; these again are made up of filaments. Every muscle and fasciculus is enveloped, and held together by a portion of cellular membrane, called its sheath, which is tough and elastic, and becomes more delicate the smaller the fasciculus or filament it surrounds.

The uses of the sheaths are to facilitate the sliding of the muscles upon each other, and by their pressure to add to their strength and tenacity. Towards their extremities the muscles are attached to a white, shining, fibrous structure termed tendons, which serve to fasten them securely to the surfaces of bones; the extremity, which is the more fixed, is termed the head, or *origin*, and that which is the more movable, the tail, or *insertion;* and the middle is the *belly*, or swell. The fibres of the most simple muscles are longitudinal. In others the fibres are disposed in rays which converge to a tendinous point, hence called *radiate* muscles; others again have their fibres converging to one or both sides of a tendon like the plumes of a pen; these are called the *penniform*, and the *bi-penniform*.

There are two classes of muscles, the voluntary, or those which are influenced by the will, and the involuntary, or those not governed by the will. The first class, which is much the most numerous, are placed between the skeleton and the skin, and constitute the chief bulk of the body; the second class are contained within the great cavities of the skeleton, and enter into the formation of the digestive, circulatory, and urinary organs, performing most of the internal movements of the animal economy. The color of the muscles of voluntary motion is of a decided red, that of the involuntary muscles is lighter; the fibres of the latter, also, cross each other and interlace.

The muscles are divided into four parts, viz. those of the head and neck; those of the trunk; those of the upper extremities; and those of the lower extremities.

Muscles of the Head and Neck.

First, of the Face.

The *occipito frontalis* is a thin muscle passing from the back to the front of the head immediately beneath the scalp; it has four bellies, and arises from the superior semi-circular ridge of the occiput, and is inserted into the superior margin of the orbicularis oris, and corrugator supercilia, and into the internal angular process of the os frontis and nasi. It pulls the skin of head backwards and forwards, and elevates the eye-brows, making transverse wrinkles.

Fig. 11.

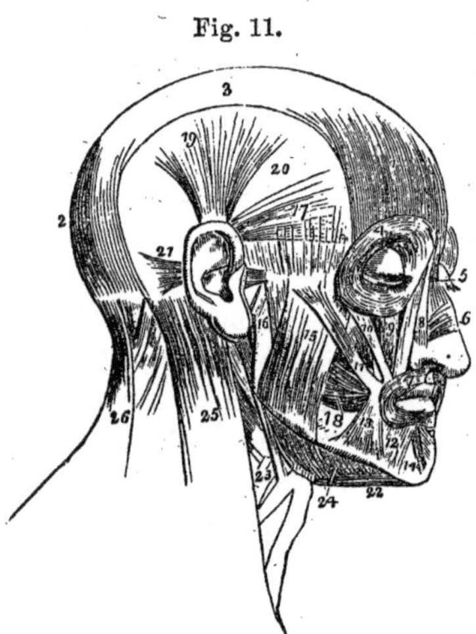

Fig. 11. View of the muscles of the head and neck. 1, 2, 3 Occipito frontalis; 4 orbicularis palpebrarum; 5 corrugator supercilii; 6 compressor naris; 7 orbicularis oris; 8 levator labii superioris alæque nasi; 9 levator anguli oris; 10 zygomaticus minor; 11 zygomaticus major; 12 depressor labii inferioris; 13 depressor anguli oris; 14 levator mentii; 15 masseter; 16 temporalis; 18 buccinator; 24 mylo-hyoideus; 25 sterno-cleideo-mastoideus.

The *compressor naris* arises from the root of the ala nasi, and is inserted into its fellow on the dorsum of the nose, and into the lower part of the os nasi. It serve both to compress and dilate the nostril.

The *orbicularis palpebrarum* is a sphincter surrounding the orbits of the eyes; it arises from the nasal process of the superior maxillary bone, the os unguis, the internal angular process of the os frontis, and the superior margin of the internal palpebral ligament; and is inserted into the orbital and nasal processes of the superior maxillary, and into the inferior margin of the palpebral ligament. It closes the eyes.

The *corrugator supercilii* arises from the internal angular process of the os frontis, and is inserted into the occipito frontalis and orbicularis. It is small and pointed, and makes the vertical wrinkles of the forehead.

The *levator labii superioris alæque nasi* is placed at the side of the nose; it arises from the nasal and orbitar processes of the superior maxillary bone, and is inserted into the upper lip and the wing of the nose. It draws the upper lip and wing of the nose upwards.

The *levator anguli oris* is small and concealed by the last; it arises from the superior maxillary bone below the infra-orbital foramen, and is inserted into the angle of the mouth. Its use is to elevate the angle of the mouth.

The *zygomaticus minor* arises from the fore part of the malar bone, and is inserted into the upper lip. It draws the corner of the mouth obliquely upwards and outwards.

The *zygomaticus major* arises from the malar bone outside of the last, and is inserted into the corner of the mouth. Its use same as the last.

The *depressor labii superioris alæque nasi* arises from the alveolar processes of the incisor and canine teeth, and is inserted into the wing of the nose and upper lip. It depresses the upper lip and the wing of the nose.

The *depressor anguli oris* arises from the base of the lower jaw at the side of the chin, and is inserted into the corner of the mouth. It draws the corner of the mouth downwards.

The *depressor labii inferioris* arises from the base of the lower jaw, beneath the last, and is inserted into the whole side of the lower lip. It draws the lip downwards.

The *levator menti*, or *labii inferioris* arises from the alveolar processes of the lateral incisor and canine teeth, and is inserted into the lower lip. It elevates the lip.

The *buccinator* arises from the coronoid process of the lower jaw, the tuber of the upper jaw, and the alveolar processes of both jaws, and is inserted into the corner of the mouth and lips. It draws the corner of the mouth backwards.

The *orbicularis oris* is a circular muscle, which surrounds the mouth, constituting the principal part of the lips; it has no long origin but is attached to the surrounding muscles, and acts as the antagonist to most of them.

The *masseter* arises from the upper maxillary, malar, and the zygomatic process of the temporal bones, and is inserted into the angle and outer surface of the lower jaw. It closes the jaws, and also draws the lower jaw backwards and forwards.

The *temporalis* arises from the temporal fascia, and from the sides of the temporal, frontal, and parietal bones, and is inserted into the coronoid process of the lower jaw. It pulls the lower jaw upwards.

The *pterygoideus externus* arises from the pterygoid, spinous, and temporal processes of the sphenoid bone, and from the tuber of the upper maxillary, and is inserted into the neck of the lower jaw. It draws the lower jaw upwards.

The *pterygoideus internus* arises from the pterygoid fossa, Eustachian tube, and internal pterygoid process, and is inserted into the inner surface of the neck of the lower jaw. It draws the jaw upwards.

Second, of the Neck.

There are two layers of cellular tissue in the neck, called the superficial, and the deep seated *fascia*; the first is immediately beneath the skin, and a continuation of the fascia covering the whole body; the second is more condensed, and extends from the ligament nuchæ to the sternum, investing the several muscles and blood-vessels.

The *platisma myoides* is a broad thin muscle, placed between two laminæ of the superficial fascia, and is attached to the cellular tissue below the clavicle, and to the muscles at the side of the face and lower jaw. It elevates the skin of the neck.

The *sterno-cleido-mastoideus* arises from the upper part of the sternum, and the sternal end of the clavicle, and passing obliquely across the neck is inserted into the mastoid process, and the adjoining ridge of the occipital bone. It forms a prominent ridge on the outside of the neck. Its use is to draw the chin to the sternum.

The *sterno-hyoideus* arises from the sternum, clavicle, and cartilage of the first rib, and is inserted into the lower edge of the hyoid bone. It draws the hyoid bone towards the sternum.

The *sterno-thyoideus* arises from the sternum and cartilage of the first rib, and is inserted into the side of the thyroid cartilage. It draws this cartilage downwards.

The *thyro-hyoideus* arises from the side of the thyroid cartilage, and is inserted into the base and cornu of the os hyoideus. It draws the os hyoideus and thyroid cartilage together.

The *omo-hyoideus* arises from the upper edge of the scapula, and is inserted into the base of the os hyoideus. It draws the os hyoideus downwards.

The *digastricus* arises from the fossa at the base of the mastoid process of the temporal bone, and is inserted into the base of the lower jaw at the side of its symphysis; its middle is tendinous, and passes through the stylo-hyoid muscle. Its use is to raise the hyoid bone, and open the mouth, even when the lower jaw is fixed.

The *stylo-hyoideus* arises from the middle and lower part of the styloid process of the temporal bone, and is inserted into the base and cornu of the os hyoides. It draws the os hyoides upwards and backwards.

The *stylo-glossus* arises from the upper and internal part of the styloid process, and is inserted into the side of the root of the tongue. It draws the tongue backwards.

The *stylo-pharyngeus* arises from the inner side of the root of the styloid process, and is inserted into the side of the pharynx between the superior and middle constrictors, and is inserted into the posterior margin of the thyroid cartilage. It draws the larynx and pharynx upwards.

The *mylo-hyoideus* arises from the root of the alveolar process of the lower jaw, and is inserted into a white tendinous line, placed between it and its fellow. It forms the floor of the mouth. Its use is to draw the os hyoides upwards and project the tongue.

The *genio-hyoideus* is placed beneath the last, and arising from the tubercle on the posterior side of the chin, is inserted into the body of the hyoid bone.

The following seven pairs of muscles are placed at the front and sides of the cervical vertebræ.

The *longus colli* arises from the three upper dorsal vertebræ, and the transverse processes of the five lower cervical vertebræ; and is inserted into the bodies of all the cervical vertebræ. It bends the neck forwards and to one side.

The *rectus capitis anticus major* arises from the transverse processes of the third, fourth, and fifth cervical vertebræ, and is inserted into the cuneiform process of the os occipitus. It bends the head forwards.

The *rectus capitis anticus minor* arises from the front of the first cervical vertebræ, and is inserted into the basilar process of the occiput. It bends the head forwards.

The *rectus capitis lateralis* arises from the front of the transverse process of the atlas, and is inserted into the ridge leading from the condyle to the mastoid process of the occiput. It draws the head slightly to one side.

The *scalenus anticus* arises from the transverse processes of the fourth, fifth, and sixth cervical vertebræ, and is inserted into the upper surface of the first rib near its middle. It bends the neck forwards, or elevates the first rib.

The *scalenus medius* arises from the transverse processes of all the cervical vertebræ, and is inserted into the upper surface of the first rib from the middle to the tubercle. Its use same as the last.

The *scalenus posticus* arises from the transverse processes of the fifth and sixth cervical vertebræ, and is inserted into the upper face of the second rib, beyond its tubercle. It bends the neck, and raises the second rib. The use of the three scalenii muscles, acting together, is to elevate the ribs, and bend the neck to one side.

Muscles of the Trunk.

First, those of the front of the Thorax.

The *pectoralis major* forms the large fleshy cushion of the chest. It arises from the first two bones of the sternum, the cartilages of the fifth and sixth ribs, the anterior two-thirds of the clavicle, and the tendon of the external oblique muscle, and is inserted by a broad thin tendon into the outer edge of the bicipital groove of the humerus. It draws the arm inwards and forwards, and depresses it when raised.

The *pectoralis minor* is beneath the last; it arises from the upper edges of the third, fourth, and fifth ribs, and is inserted into the inner faces of the coracoid process of the scapula. It draws the scapula inwards and downwards.

The *subclavius* arises from the cartilage of the first rib, and is inserted into the lower face of the clavicle. It is placed immediately beneath the clavicle, and draws it downwards.

The *serratus magnus*, or *serratus major anticus* is a broad muscle at the sides of the ribs; it arises from the nine upper ribs by fleshy digitations, the five lower of which are connected with the internal oblique muscle, and is inserted into the base of the scapula. It draws the scapula forwards.

The *intercostales* fill up the spaces—eleven in number—between the ribs. There are two to each space, an external, and an internal. The external arises from the transverse process of the vertebræ, and the inferior acute edge of the rib, and is inserted into the superior rounded edge of the rib below; its fibres passing obliquely forwards and downwards. The internal arises from the inferior edge of the rib, and is inserted into the superior rounded edge of the rib below; its fibres pass obliquely backwards and downwards. The use of the intercostales is to draw the ribs together.

The *triangularis sterni* is on the posterior surface of the cartilages of the ribs; it arises from the ensiform cartilage and the second bone of the sternum, and is inserted into the cartilage of the third, fourth, fifth, and sixth ribs. Its use is to depress the ribs, and thereby diminish the cavity of the thorax.

Fig. 12.

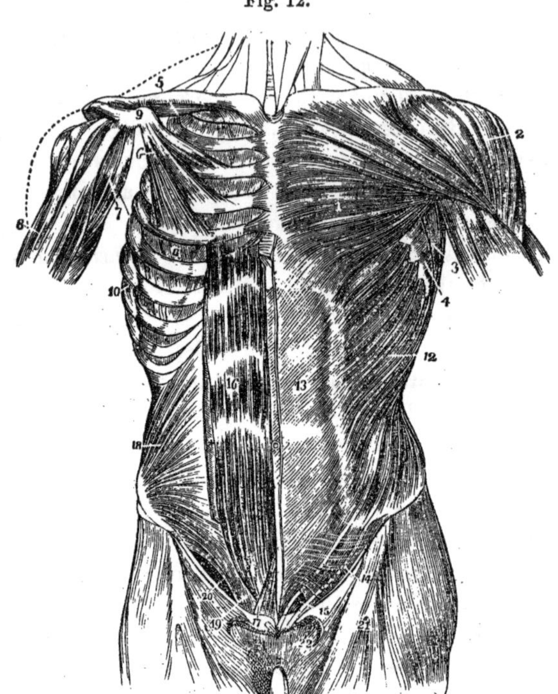

Fig. 12. View of the muscles in front of the thorax and abdomen. 1 Pectoralis major; 2 deltoid; 3 latissimus dorsi; 4 origin of the serratus magnus; 5 subclavius; 6 pectoralis minor; 7 coraco-brachialis; 8 biceps; 9 coracoid process of the scapula; 10 serratus magnus; 11 external intercostal muscle; 12 external oblique; 13 aponeurosis of the external oblique; 14 and 15 Poupart's ligament; 16 rectus; 17 pyramidalis; 18 internal oblique; 19 tendon of the internal oblique and transversalis; 20 passage of the spermatic cord.

Second, muscles of the abdomen.

Between the skin and the muscles of the abdomen is a layer of condensed cellular substance, called the *fascia superficialis abdominis:* it may be traced upwards over the thorax to the neck and face, and downwards to the thigh. It is generally blended with fat, and in the groin encloses the lymphatic glands, and the external pudic vessels between its laminæ.

There are five pairs of abdominal muscles, viz. the external oblique, internal oblique, rectus abdominis, pyramidalis, and transversalis.

The *obliquus externus* arises from the eight inferior ribs, by tendinous and fleshy heads, the five upper of which are interlocked with those of the serratus anticus, and the three lower with those of the latissimus dorsi; its fibres pass obliquely downwards and forwards, and are inserted into the whole length of the linea alba, and into the anterior half or two-thirds of the crest of the ilium: the tendon also extends anteriorly to the body and symphysis of the pubes, forming thereby Poupart's ligament.

The *linea alba,* in the middle line of the body, is formed by the union of the tendons of the three broad muscles on either side of the abdomen. On either side of the linea alba, and two or three inches from it, is another line formed by the same tendons, called *linea semilunaris.* The use of the external oblique is to compress the viscera of the abdomen, and approximate the pelvis and thorax.

The *obliquus internus* lies beneath the last, and its fibres pass in an opposite direction; it arises from the three inferior spinous processes of the lumbar vertebræ, all those of the sacrum, the crista of the ilium, and the upper half of Poupart's ligament; and is inserted into the cartilages of the six inferior ribs, the ensiform cartilage, and into the whole length of the linea alba. At the linea alba the tendon is divided into two laminæ, which enclose the rectus muscle. Its use is the same as that of the internal oblique.

The *transversalis abdominis* is underneath the last, and arises from the transverse processes of the last dorsal and four upper lumbar vertebræ, the crest of the ilium, the outer half of Poupart's ligament, and from the cartilages of the six or seven lower ribs; and is inserted into the ensiform cartilage, and linea alba. At the pubes it is inserted in common with the tendon of the internal oblique behind the external abdominal ring. Just above this insertion it is divided into two lamina to enclose the pyramidalis muscle. It compresses the viscera of the abdomen.

The *rectis abdominis* arises from the symphysis and body of the pubes, and passing upwards between the layers of the tendon of the external oblique, increasing in breadth as it ascends, is inserted into the ensiform cartilage, and the fifth, sixth and seventh ribs. It has several tendinous intersections crossing it, called *linea transversæ.* Its use is to draw the thorax towards the abdomen.

The *pyramidalis* is placed at the lower and front part of the rectus, and is about three inches long; it arises from the body of the pubes, and is inserted into the linea alba. Sometimes it is wanting. It gives strength to the lower part of the abdomen.

The *cremaster* is commonly considered as a portion of the internal oblique, as previous to the descent of the testicle it forms the lower edge of this muscle. It envelops the testicle and cord, and is inserted into the tendons of the internal oblique and transversalis. It draws up the testicle.

The *fascia transversalis* is placed between the abdominal muscles and the peritoneum; it is thin and tough, and has an opening in it, to transmit the spermatic cord, which is called the internal abdominal ring.

Third, muscles of the upper and posterior part of the abdomen.

The *diaphragm* is a movable, muscular septum, or division, placed between the cavities of the thorax and abdomen. Above, it is in contact with the pericardium and lungs, and below, with the liver, stomach, and spleen. It consist of two portions, called the greater and lesser muscle. The greater arises from the ensiform cartilage, and the six inferior ribs on each side, its fibres converge towards the centre to be inserted into the cordiform tendon. The lesser has two bellies, called the crura, the one on the right side being the larger; it arises from the second, third, and fourth lumbar vertebræ, and is also inserted into the cordiform tendon. This tendon, so called from its resemblance to a heart, is large and shining, and nearly horizontal in the erect posture; it has its apex toward the sternum, and its notch towards the spine; its summit is about on a line with the lower end of the second bone of the sternum.

There are three openings or foramina in the diaphragm; the first, situated in the back part of the muscle near the spine, is called the foramem œsophageum. It is elliptical in shape, and transmits the œsophagus and par vagum nerves. The second, called the foramen quadratum, is in the centre of the cordiform tendon and transmits the ascending vena cava. The third, the hiatus aorticus, is in front of the vertebræ, and between the cruræ; it gives passage to the aorta, thoracic duct, and great splanchnic nerves.

The *quadratus lumborum* arises from the crista of the ilium, and is inserted into the last dorsal and all the lumbar vertebræ, and into the lower edge of the last rib. It bends the loins to one side, and pulls down the last rib.

The *psoas parvus* arises from the last dorsal and the first lumbar vertebræ, and is inserted into the linea innominata and fascia iliaca. It draws up the sheath of the femoral vessels. Sometimes it is absent.

The *psoas magnus* arises from the transverse processes of all the lumbar vertebræ, and from the bodies of the four upper lumbar and the last dorsal vertebræ; and is inserted into the lesser trochanter of the femur, and for about an inch below it.

The *iliacus internus* arises from the transverse process of the last lumbar vertebræ, the costa and crest of the ilium, and from the capsula of the hip joint; and is inserted into the tendon of the psoas magnus. It assists the latter muscle in bending the body forwards and drawing the thigh upwards.

Fourth, muscles of the back.

The *trapezius* is a broad muscle immediately under the skin of the back; it arises from the occipital protuberance and the superior semicircular ridge of the occiput, from the spinous processes of the neck by the ligamentum nuchæ, and from all those of the back; and is inserted into the external third of the clavicle, and the acromion process and spine of the scapula. It draws the scapula towards the spine.

The *latissimus dorsi* covers the whole of the lower part of the back; it arises from the seven inferior spinous processes of the back, from all those of the loins and sacrum, and from three or four of the last ribs; and is inserted into the posterior ridge of the bicipital groove along with the teres minor. It draws the humerus downwards and backwards.

The *serratus inferior posticus* arises from the two inferior spinous processes of the back, and the three superior of the loins, and is inserted into the four inferior ribs. It draws the ribs downwards.

The *rhomboideus minor* arises from the three inferior spinous processes of the neck; and is inserted into the base of the scapula opposite the spine.

The *rhomboideus major* arises from the last spinous process of the neck, and the four superior of the back; and is inserted into the base of the scapula below the spine. These two muscles draw the scapula upwards and backwards.

The *serratus superior posticus* arises from the three inferior spinous processes of the neck, and the two superior of the back; and is inserted into the second, third, fourth and fifth ribs. It draws the ribs upwards.

The *levator scapulæ* arises from the three, four, or five superior transverse processes of the neck; and is inserted into the angle of the scapula and its base above the spine. It draws the scapula upwards.

The *splenius* arises from the spinous processes of the five inferior cervical and the four superior dorsal vertebræ; and is inserted into the occipital bone between the semicircular ridges, and into the transverse processes of the two superior cervical vertebræ. The part which goes to the head is called splenius capitus, and that which goes to the neck, splenius colli. It draws the head and neck backwards.

Fig. 13.

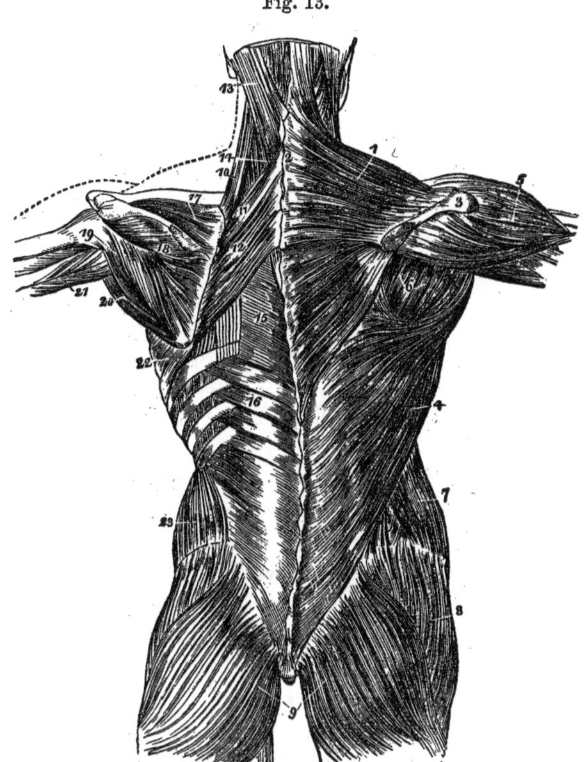

Fig. 13. View of the muscles of the back. 1, 2 The trapezius; 3 acromion process and spine of scapula; 4 latissimus dorsi; 5 deltoid; 6 infra-spinatus, and teres minor and major; 7 external oblique; 8 gluteus medius; 9 glutei maximi; 10 levator anguli scapulæ; 11 & 12 rhomboideus minor and major; 13 splenius capitis; 14 splenius colli; 15 vertebral aponuerosis 16 serratus posticus inferior; 17 supra spinatus; 18 infra spinatus; 19 teres minor; 20 teres major; 21 long head of the triceps; 22 serratus magnus; 23 internal oblique.

The *sacro-lumbalis and longissimus dorsi* arise in common from the spinous and transverse processes of the loins and sacrum, and from the crest of the ilium; and the first is inserted into the angles of the ribs. The latter, which is nearest the spine, is inserted into all the transverse processes of the vertebræ of the back except the first, and into the under-edges of all the ribs beyond their tubercles, except the two last. They keep the spine erect, and draw down the ribs.

The *spinalis dorsi* arises from the three lower spinous processes of the back, and the two upper of the loins; and is inserted into the nine upper spinous processes of the back except the first; it is almost entirely tendinous. Its use is to keep the spine erect.

The *cervicalis descendens* arises from the four superior ribs, and is inserted into the fourth, fifth, and sixth transverse processes of the back. It draws the neck backwards.

The *transversalis cervicis* arises from the five superior transverse processes of the back; and is inserted into the five middle transverse processes of the neck. It draws the head and neck backwards.

The *trachelo-mastoid* arises from the three upper transverse processes of the back, and the five inferior of the neck; and is inserted into the mastoid process.

The *complexus* arises from transverse processes of the four inferior cervical, and of the superior dorsal vertebræ; and is inserted into the os occipitus between its semicircular ridges. It draws the head backwards.

The *semi-spinalis cervicis* arises from the transverse processes of the six upper dorsal vertebræ; and is inserted into the spinous processes of the five middle cervical vertebræ. It draws the neck obliquely backwards.

The *semi-spinalis dorsi* arises from the transverse processes of the seventh, eighth, ninth, and tenth dorsal vertebræ, and is inserted into the spinous processes of the two lower cervical, and five upper dorsal vertebræ. It draws the spine obliquely backwards.

The *multifidus spinæ* arises from the oblique and transverse processes of all the sacral, lumber and dorsal vertebræ, and is inserted into the spinous processes of all the vertebræ of the loins and back, and of the four inferior of the neck. It twists the spine backwards, and keeps it erect.

Fig. 14. Posterior views of the neck, thorax, abdomen, and pelvis. 1, 2, 3 Sacro-lumbalis, and longissimus dorsi; 4 spinalis dorsi; 5 cervicalis descendens; 6 transversalis cervicis; 7 trachelo mastoid; 8 complexus; 10 semi-spinalis dorsi; 11 semi-spinalis cervicis; 12 rectus capitis posticus major; 13 rectus capitis posticus minor; 14 obliquus superior; 15 obliquus inferior; 16 multifidus spinæ; 17 levatores costarum; 18 inter-transversalis.

Fig. 14.

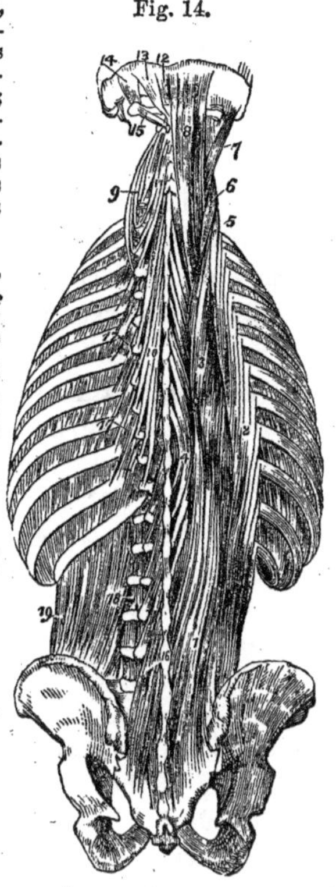

The *rectus capitis posticus major* arises from the spinous process of the dentata, and is inserted into the inferior semicircular ridge of the occiput and a part of the surface below it. It turns the head, and draws it backwards.

The *rectus capitis posticus minor* arises from the tubercle of the atlas, and is inserted into the occiput at, and below the inferior semicircular ridge. It draws the head backwards.

The *obliquus superior* arises from the transverse process of the atlas, and is inserted into the outer end of the inferior semicircular ridge of the occiput. It draws the head backwards.

The *obliquus inferior* arises from the spinous process of the dentata, and is inserted into the transverse process of the first cervical vertebræ. It rotates the first vertebræ on the second.

The *inter-spinales* are short muscles, placed between the spinous processes of all the vertebræ. They are double in the neck; tendinous in the back; and single and well marked in the loins. They keep the spine erect.

The *inter-transversalis* are placed between all the transverse processes. They draw these processes together.

The *levatores costarum.* There are twelve of these small muscles on either side of the spine; they arise from the transverse processes of the last cervical, and the eleven superior dorsal vertebræ, and are inserted into the upper edge of the two next ribs below. They elevate the ribs.

Muscles of the upper extremities.

The muscles of the upper extremity are covered with the brachial fascia, which extends from the shoulder to the hand. It forms at the wrist the anterior and posterior annular ligaments, which holds down the flexor tendons of the hand and fingers. In the hand it forms the palmar aponeurosis.

First, muscles of the shoulder.

The *deltoid* passes over the top of the shoulder, giving to it its rotundity: it arises from the spine of the scapula, the acromium process, and the outer third of the clavicle, and is inserted into the triangular roughness near the middle of the humerus. It raises the humerus.

The *supra-spinatus scapulæ* arises from the whole of the fossa supra-spinata, and is inserted into the inner facet of the greater tuberosity of the humerus. It raises the arm, and turns it outwards.

The *infra-spinatus scapulæ* arises from the whole of the fossa infra-spinata, and is inserted into the middle facet of the greater tuberosity of the humerus. It rolls the humerus outwards and backwards.

The *teres minor* arises from the lesser costa of the scapula, and is inserted into the outer facet of the greater tuberosity of the humerus. It rotates the humerus outwards, and draws it downwards and backwards.

The *teres major* arises from the posterior surface of the angle of the scapula, and from a part of its inferior costa; and is inserted into the posterior ridge of the bicipital groove along with the tendon of the latissimus dorsi.

The *subscapularis* arises from the whole of the inferior or thoracic surface of the scapula, and is inserted into the lesser tuberosity of the humerus. It rotates the humerus inwards and draws it downwards.

Second, muscles of the arm.

The *biceps flexor cubiti* has two heads, one the long head, arising from the superior extremity of the glenoid cavity of the scapula, and passing through the joint and bicipital groove; the other, called the short head, arising from the coracoid process of the scapula. It is inserted into the posterior rough parts of the tubercle of the radius. Its use is to flex the fore-arm.

The *coraco-brachialis* arises in common with the short head of the biceps from the middle facet of the coracoid process of the scapula, and is inserted by a rough ridge into the internal side of the middle of the humerus. It draws the arm upwards and inwards.

The *brachialis internus* arises from the anterior and lower half of the humerus, and is inserted into the rough surface at the root of the coronoid process of the ulna. It flexes the fore-arm.

The *triceps extensor cubicti* arises by three heads, the first, called the longus, from a rough ridge on the inferior edge of the scapula; the second, the brevis, from a ridge on the back part of the humerus just below the head; and the third, the brachialis internus, from the inner side of the humerus near the insertion of the teres major. It is inserted into the back-part of the olecranon process. It extends the fore-arm.

The *anconeus* arises from the external condyle of the humerus, and is inserted into the ulna below the olecranon process. It extends the fore-arm.

Third, muscles of the fore-arm.

There are eighteen of these; eight on the front of the fore-arm, which are the flexors, and ten on the back part, which are the extensors. The last arise principally from the external condyle.

On the front are:

The *pronator radii teres,* arising from the internal condyle of the humerus, and the coronoid process of the ulna; and is inserted into the middle and back part of the radius.

The *flexor carpi radialis*, arising from the internal condyle of the humerus, the adjacent fascia, and from the upper part of the ulna, and is inserted into the base of the metacarpal bone of the fore-finger. It flexes the hand.

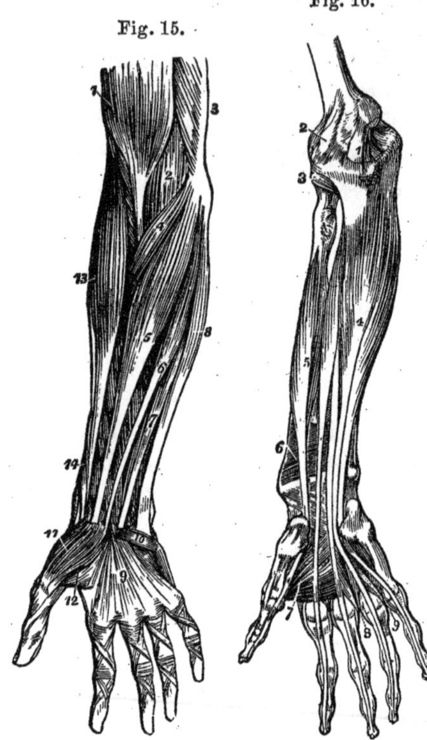

Fig. 15. View of the superficial muscles on the front of the fore-arm. 1 Lower part of the biceps; 2 part of the brachialis anticus; 3 part of the triceps; 4 pronator radii teres; 5 flexor carpi radialis; 6 palmarus longus; 7 part of the flexor sublimus digitorum; 8 flexor carpi ulnaris; 9 palmar fascia; 10 palmaris brevis muscle; 11 abductor pollicis; 12 part of the flexor brevis pollicis; 13 supinator longus; 14 extensor ossis metacarpi, and extensor primi internodii pollicis, curving around the edge of the fore-arm.

Fig. 16. View of the deep seated muscles on the front of the fore-arm. 1 The internal lateral ligament of the elbow-joint; 2 anterior ligament; 3 orbicular ligament of the head of the radius; 4 flexor profundis digitorum muscle; 5 flexor longus pollicis; 6 pronator quadratus; 7 adductor pollicis; 8 dorsal interosseous muscle of the middle finger and palmar interrosseous of the ring finger; 9 dorsal interrosseus muscle of the ring finger and palmar interosseous of the little finger.

The *flexor carpi ulnaris*, arising from the internal condyle and from the ridge at the internal side of the ulna; and is inserted into the os pisiforme, and sometimes also into the base of the metacarpal bone of the little finger. It bends the hand, and draws it towards the ulna.

The *flexor sublimis digitorum perforatus* arises from the internal condyle of the humerus, the coronoid process of the ulna, and the tubercle of the radius; and is inserted by four tendons into the second phalanges of the fingers. It bends the hand and fingers.

The *flexor profundus digitorum perforans* arises from the anterior flat surface of the ulna, from the coronoid process, and the interosseous ligament; and is inserted by four tendons, which pass through the slit in those of the flexor sublimis, into the third phalanges of the fingers. It bends the last phalanges of the fingers, and also the hand.

The *flexor longus pollicis* arises from the middle two-thirds of the radius, the interosseous ligament, and the internal condyle of the humerus; and is inserted into the base of the second phalanx of the thumb. It bends the last joint of the thumb.

The *pronator quadratus* arises from the inner surface of the ulna, near its lower extremity, and passing obliquely across the fore-arm, is inserted into the corresponding surface of the radius. It rotates the radius inwards.

On the back of the fore-arm are:

The *supinator radii longus*, arising from the ridge leading to the external condyle; and is inserted into a rough ridge just above the styloid process of the radius. It rotates the radius outwards.

The *extensor carpi radialis longus* arising from the ridge of the external condyle of the humerus; and is inserted into the posterior part of the root of the metacarpal bone of the forefinger. It extends the hand.

The *extensor carpi radialis brevis*, arising from the external condyle of the humerus, and from the external lateral ligament; and is inserted into the posterior part of the base of the metacarpal bone of the second finger. It extends the hand.

The *extensor carpi ulnaris*, arising from the external condyle and from the fascia, is inserted into the ulnar side of the base of the metacarpal bone of the little finger. It extends the hand.

Fig. 17.

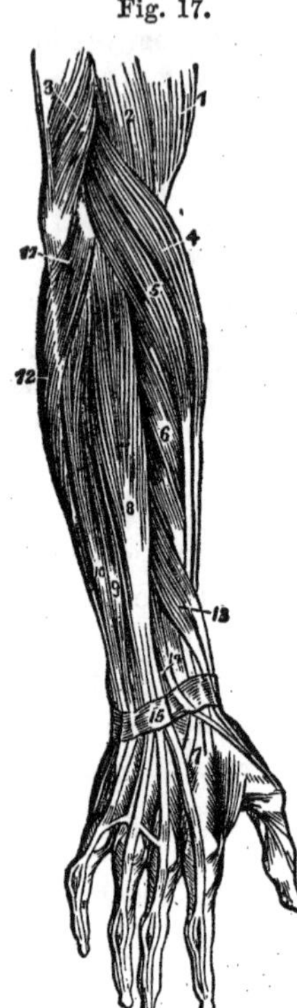

Fig. 17. View of the muscles on the back of the fore-arm. 1 Lower part of biceps; 2 part of the brachialis anticus; 3 insertion of the triceps into the olecranon; 4 supinator longus; 5 extensor carpi radialis longior; 6 extensor carpi radialis brevior; 7 tendons showing the insertion of the two last muscles; 8 extensor communis digitorum; 9 extensor minimi digiti; 10 extensor carpi ulnaris; 11 anconeus; 12 part of the flexor carpi ulnaris; 13 extensor ossis metacarpi and extensor primi internodii; 14 extensor secundi internodii; 15 posterior annular ligament, beneath which are the tendons of the extensor communis.

The *extensor digitorum communis*, arising from the external condyle; and is inserted by four tendons, which are connected by slips, near the roots of the fingers, into all the phalanges of the fingers. It extends the joints of the fingers.

The *supinator radii brevis*, arising from the external condyle and from the ridge on the posterior radial edge of the ulna; and is inserted into the tubercle and the oblique rough ridge of the radius. It rotates the radius outwards.

The *extensor ossi metacarpi pollicis manus*, arising from the posterior part of the ulna, the interosseous ligament, and from the back part of the radius; and is inserted into the base of the metacarpal bone of the thumb, and into the trapezium. It extends the metacarpal bone of the thumb.

The *extensor minor pollicis manus*, arising from the back of the ulna below its middle and from the interosseous ligament; and is inserted into the first phalanx of the thumb. It extends the first phalanx.

The *extensor major pollicis manus*, arising from the back of the ulna above its middle, the interosseous ligament and the back of the radius; and is inserted into the base of the second phalanx of the thumb. It extends the second phalanx.

The *indicator*, arising from the back of the ulna and from the interosseous ligament; and is inserted into the back of the fore-finger, as far as the base of the third phalanx. It extends the fore-finger.

Of the Muscles of the Hand.

The *palmaris brevis* is just beneath the skin at the inner side of the hand; it arises from the anterior ligament of the wrist, and from the palmar aponeurosis; and is inserted into the skin at the inner margin of the hand. It contracts the skin of the hand.

The *lumbricales* are four small muscles, resembling earth-worms, which arise from the tendons of the flexor profundus, and are inserted into the radial side of the first phalanx of each finger. They bend the first phalanges.

The *abductor pollicis manus* arises from the annular ligament and from the ends of the scaphoid and trapezium; and is inserted into the base of the first phalanx of the thumb. It draws the thumb from the fingers.

The *opponens pollicis* arises from the trapezium and the annular ligament; and is inserted into the radial edge of the metacarpal bone of the thumb. It draws the metacarpal bone inwards.

The *flexor brevis pollicis manus* has two bellies, the first head arises from the trapezium, trapezoides, and annular ligament, and is inserted into the outer side of the first phalanx of the thumb; the second head arises from the magnum, unciforme, and the base of the metacarpal bone of the middle finger, and is inserted into the inner side of the base of the first phalanx of the thumb. The sesamoid bones are included in these tendons. It bends the first phalanx to the thumb.

The *adductor pollicis manus* arises from the ulnar margin of the metacarpal bone of the middle-finger, and is inserted into the inner side of the base of the first phalanx of the thumb. It draws the thumb towards the fingers.

Fig. 18.

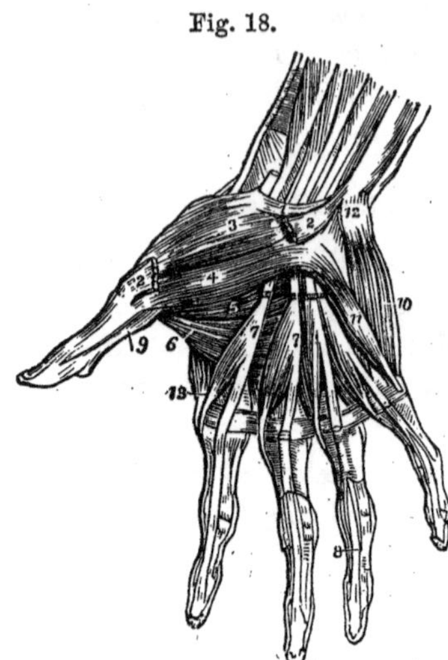

Fig. 18. View of the muscles of the palm of the hand. 2 Abductor pollicis manus; 3 opponens pollicis; 4, 5 flexor brevis pollicis manus; 6 adductor pollicis manus; 77 lumbricales; 10 abductor minimi digiti manus; 11 flexor parvus minimi digiti.

The following three muscles are situated at the ulnar side of the hand, constituting the ball of that side.

The *abductor minimi digiti manus* arises from the pisiform bone and annular ligament, and is inserted into the ulnar side of the first phalanx of the little finger. It draws the little finger from the rest.

The *flexor parvus minimi digiti manus* arises from the unciform bone and annular ligament, and is inserted into the ulnar side of the base of the first phalanx of the little finger. It bends the little finger.

The *adductor metacarpi minimi digiti* arises from the unciform process and annular ligament, and is inserted into the metacarpal bone of the little finger along its whole length. It brings the metacarpal bone towards the wrist.

There are seven interosseous muscles, which fill up the interstices of the metacarpal bones. Three of them are adductors and placed on the palmar side; and four are abductors placed on the dorsal side.

The *adductors* arise from the base of the metacarpal bone of one finger, beginning at the index, and are inserted into the base of the first phalanx of the same finger.

The *adductors* are penniform, and arise by two heads from the adjoining sides of the metacarpal bones, and are inserted into the bases of the first phalanges; two of them, the second and third, go to the middle finger.

Of the Muscles of the lower extremities.

Like the rest of the body the lower extremities are invested by cellular membrane, reaching from the crest of the ilium to the foot. It has received the general name of *fascia lata*, and is exceedingly strong and dense. That portion in front of the thigh is termed *iliac* and *pubic;* that which surrounds the knee *involucrum;* that of the leg *crural*; and that in the sole of the foot *plantar fascia*. In front of the ankle it forms the annular ligament, and under the sinuosity of the os calcis, where it binds down the flexor tendon, the ligamentum lancinatum. It also furnishes sheaths for the muscles.

The *tensor vaginæ femoris* muscle arises from the anterior superior spinous process of the ilium; and is inserted into the fascia of the thigh. It makes the fascia tense, and rotates the foot inwards.

The *sartorius* arises from the anterior superior spinous process of the ilium; and is inserted into the inner side of the head of the tibia. It is the longest muscle in the body. It bends the leg, and draws it inwards.

The *rectus femoris* arises from the anterior inferior spinous process of the ilium, and is inserted into the upper surface of the patella, and through the ligament of the patella into the head of the tibia. It extends the leg.

The *vastus externus* arises from the trochanter major and linea aspera, and is inserted into the outer and upper part of the patella. It extends the leg.

The *vastus internus* arises from the whole length of the linea aspera, and covering the entire inside of the thigh is inserted into the internal edge of the tendon of the rectus, and into the external and upper part of the patella. It likewise extends the leg.

The *cruræus* arises from the front of the thigh and from the linea aspera; and is inserted into the upper surface of the patella. It extends the leg.

The four last named muscles form a common tendon (the ligamentum patellæ), in which is included the patella, before it is inserted into the head of the tibia. In consequence of this arrangement, and of these muscles, having a common use, they are often described as the *quadriceps femoris.*

The *gracilis* arises from the front and ramus of the pubes, and is inserted into the inside of the head of the tibia. It flexes the leg.

The *pectineus* arises from the upper face and front of the pubes, and is inserted into the linea aspera below the trochanter minor. It draws the thigh inwards and forwards.

The *adductor longus* arises from the upper front part of the pubes, and is inserted into the inner edge of the middle third of the linea aspera.

The *adductor brevis* arises from the body and ramus of the pubes, and is inserted into the inner edge of the linea aspera, along its upper third.

The *adductor magnus* arises from the body of the pubes, and the ramus of the pubis and ischium; and is inserted into the whole length of the linea aspera. These three adductors have the same use, that of drawing the thigh forwards and inwards.

The *gluteus magnus* or *maximus* arises from the crista of the ilium, the sides of the sacrum and coccyx, and from the great sacro-sciatic ligament; and is inserted into the upper part of the linea aspera, and the fascia of the thigh. It draws the thigh backwards, and keeps the trunk erect.

The *gluteus medius* arises from the crest of the ilium, its dorsum between the crest and semicircular ridge, from the space between the anterior spinous processes, and from the fascia femoris; and is inserted into the upper surface of the trochanter major and the shaft of the bone in front of it. It draws the thigh backwards and outwards.

The *gluteus minimus* arises from the dorsum of the ilium between the semicircular ridge and capsular ligament; and is inserted into the upper part of the trochanter major. It abducts the thigh, and also rotates it inwards.—The three gluteii muscles form most of the fleshy part of the hip.

The *pyriformis* arises from the anterior face of the second, third and fourth bones of the sacrum, and is inserted into the superior middle part of the trochanter major. It rotates the limb outwards.

The *Gemini* are two small muscles; the upper arises from the root of the spinous process of the ischium; the lower from the back part of the tuberosity of the ischium; they are both inserted into the root of the trochanter major. They also rotate the limb outwards.

The *obturator internus* arises from the margin of the thyroid foramen, the posterior face of the thyroid ligament, and from the iliac fascia; and is inserted into the pit on the back part of the femur at the root of the trochanter major. It rotates the limb outwards.

The *quadratus femoris* arises from the tuberosity of the ischium, and is inserted into the ridge between the two trochanters of the os femoris. It rotates the limb outwards.

The *obturator externus* arises from the anterior margin of the thyroid foramen, and from the anterior face of the thyroid ligament; and is inserted into the cavity at the root of the trochanter major of the os femoris. It rotates the thigh outwards.

The *biceps flexor cruris* forms the outer hamstring; it arises by two heads; a long one from the tuberosity of the ischium in common with the semi-tendinosus, and a short one from the lower part of the linea aspera. It is inserted into the head of the fibula. It flexes the leg on the thigh.

The *semi-tendinosus* arises from the tuberosity of the ischium, and is inserted into the side of the tibia just below its tubercle. It flexes the leg on the thigh.

The *semi-membranosus* arises from the outer and upper part of the tuberosity of the ischium; and is inserted into the inner and back part of the tibia just below the joint. It flexes the leg on the thigh.

Muscles of the Leg.

The *tibialis anticus* arises from the outer side of the head and from the spine of the tibia, and from the interosseous ligament; and is inserted, at the inner side of the sole of the foot, into the base of the internal cuneiform bone and the base of the metacarpal bone of the great toe. It bends the foot, and presents the sole obliquely outwards.

The *extensor longus digitorum pedis* arises from the head of the tibia, from the head and anterior margin of the fibula, and from the interosseous ligament; and is inserted after dividing into four tendons, into the phalanges of the toes. It extends the toes, but flexes the foot.

The *peroneus tertius* arises from the lower third of the anterior angle of the tibia; and is inserted into the base of the metatarsal bone of the little toe. It is rather a part of the extensor longus than a distinct muscle. It assists in bending the foot.

The *extensor proprius pollicis pedis* arises from the lower three-fourths of the fibula, and the interosseous ligament; and is inserted into the base of the first and second phalanx of the great toe. It extends the great toe.

The *peroneus longus* arises from the outside of the head and upper third of the fibula; and is inserted into the base of the metatarsal bone of the great toe. It extends the foot, and inclines the sole outwards.

The *peroneus brevis* arises from the lower two-thirds of the outer surface of the fibula; and is inserted into the base of the metatarsal bone of the little toe. It extends the foot and presents the sole obliquely downwards.

The *gastrocnemius* has two heads, one of whlch arises from each of the condyles of the humerus; it is inserted by the tendo Achillis into the os calcis.

The *soleus* arises from the posterior part of the upper two-thirds of the fibula, and from the middle third of the tibia; and is inserted into the tendo Achillis.

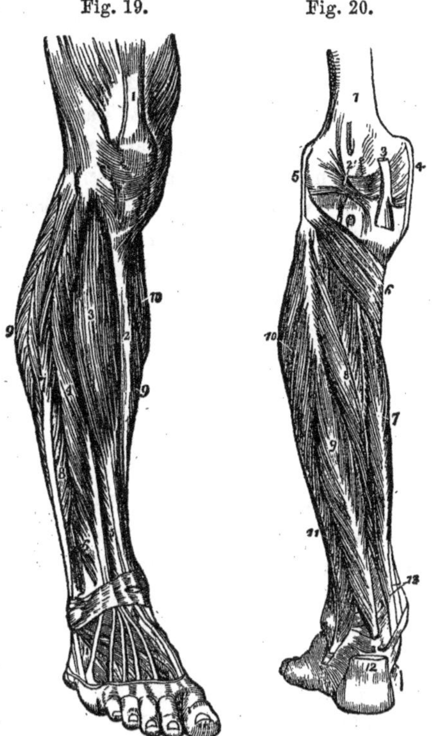

Fig. 19. View of the anterior muscles of the leg. 1 Tendon of the quadriceps femoris; 2 anterior angle of the tibia; 3 tibialis anticus; 4 extensor longus digitorum pedis; 5 extensor proprius pollicis pedis; 6 peroneus tertius; 7 peroneus longus: 8 peroneus brevis; 9, 9 borders of the soleus; 10 gastrocnemius.

Fig. 20. View of the deep seated muscles on the back of the leg. 1 Lower extremity of the femur; 2 ligament of Winslow; 3 tendon of the semi-membranosus; 4, 5 internal and external lateral ligaments; 6 popliteus muscle; 7 flexor longus digitorum; 8 tibialis profundus; 9 flexor longus pollicis; 10 peroneus longus; 11 peroneus brevis; 12 lower end of the tendo Achillis; 13 tendons of the tibialis posticus and flexor longus digitorum. To show these muscles, the gastrocnemius, plantaris and soleus muscles have been removed.

The *plantaris* arises by a very long tendon from the femur, just above the external condyle; and is inserted into the os calcis in front of the tendo Achillis.

The last three muscles have the same action, to wit, to extend the foot. They constitute the calf of the leg and are sometimes called the *triceps suræ.*

The *popliteus* arises from the external face of the external condyle, and is inserted into the oblique ridge on the back of the tibia, just below its head. It bends the leg, and rotates it in wards.

The *flexor longus digitorum pedis perforans* arises from the back of the tibia below its oblique ridge, and after dividing into four tendons is inserted into the bases of the last phalanges of the four lesser toes. These tendons perforate those of the flexor brevis. It flexes the toes, and extends the foot.

The *flexor longus pollicis pedis* arises from the posterior face of the tibia, commencing about three inches below its head, and continuing nearly to the ankle; and is inserted into the second phalanx of the great toe. It bends the great toe.

The *tibialis posticus* arises by two heads from the tibia and fibula, also from the interosseous ligament; and is inserted into the tuberosity of the scaphoid bone. It extends the foot.

Muscles of the Foot.

The *extensor brevis digitorum pedis* arises from the greater apophysis of the os calcis; and is inserted by four tendons into the backs of the four greater toes. Its tendons join those of the extensor longus. It extends the toes.

The *flexor brevis digitorum pedis* arises from the tuberosity of the os calcis and the plantar fascia, and is inserted into the second phalanges of the four smaller toes. Its tendons are perforated by those of the flexor longus. It bends the second joint of the toes.

The *flexor accessorius* arises from the inside of the sinuosity, and the front of the tuberosity of the os calcis; and is inserted into the outside of the tendon of the os calcis at its division. It assists in flexing the toes.

The *lumbricales pedis* are four small muscles, which arise from the tendon of the flexor longus; and are inserted into the inside of the first phalanx of each of the lesser toes. They assist in flexing the toes.

The *abductor pollicis pedis* arises from the internal tuberosity of the os calcis, the internal side of the navicular and internal cuneiform bones, and from the plantar fascia; and is inserted into the inner side of the base of the first phalanx of the great toe, including the internal sesamoid bone. It draws the great toe from the others.

The *flexor brevis pollicis pedis* has two bellies; it arises from the calcaneo-cuboid ligament, and the internal cuneiform bone; and is inserted by two tendons into the internal and external sesamoid bones, and into the first phalanx of the great toe.

The *adductor pollicis pedis* arises from the calcaneo-cuboid ligament, and the basis of the second, third, and fourth metatarsal bones of the lesser toes; and is inserted into the external sesamoid bone, and the tendon of the flexor brevis. It draws the great toe towards the others.

The *abductor minimi digiti pedis* arises from the outer tuberosity of the os calcis, and the metatarsal bone of the little toe; and is inserted into the base of the first phalanx of the little toe. It draws the little toe from the others.

The *flexor brevis minimi digiti pedis* arises from the calcaneo-cuboid ligament, and the fifth metatarsal bone; and is inserted into the head of the metatarsal bone, and the base of the first phalanx of the little toe. It bends the little toe.

The *transversalis pedis* is a small muscle, which lies across the anterior extremities of the metatarsal bones; it arises from the capsular ligaments of the first joints of the fourth and fifth toes; and is inserted into the external sesamoid bone. It approximates the heads of the metatarsal bones.

The *interosseus muscles* are seven in number; four of them are upon the dorsal, and three upon the plantar surface of the foot. There are two to the first smaller toe, two to the second, two to the third, and one to the fourth or little toe. The *dorsal interossei* arise by double heads from the adjacent sides of the metatarsal bones; and are inserted into the base of the first phalanx. The first is an *adductor* being inserted into the *inner* side of the second toe; the other three are inserted into the *outer* side of the second, third, and fourth toes, and are consequently abductors.

The *plantar interossei* arise from the base of the metatarsal bones of the three outer toes; and are inserted into the inner side of the base of the first phalanx of the same toes. They draw the toes inwards, or *adduct* them.

Organs of Digestion.

Mouth.—The mouth is bounded anteriorly and laterally by the lips and cheeks; its roof is formed by the hard and soft palate; and its floor by the mylo-hyoid muscles; posteriorly it extends to the pharynx, and communicates with the fauces. The space between the lips and the teeth is sometimes called the vestibule of the mouth. The mouth is lined by a mucous membrane, beneath which are a number of muciparous glands. This membrane is thrown into folds (frena) at several points, the one beneath the tongue is called frenum linguæ.

The *lips* (labia) are composed of muscular fibres and fat, and covered externally by skin. The upper is longer and thicker than the lower, and has a vertical depression on the middle front surface, called *philtrum*.

The *gums* are formed of the lining membrane of the mouth much thickened.

Tongue.—The tongue is the special organ of taste, and is also of material importance in speech and mastication. It is a muscular body, symmetrical, oblong, and flattened; its size and shape are variable.

The posterior extremity is called its base, or root, the anterior its point, or tip, and the middle its body. The root is attached to the hyoid bone.

The mucous covering is very thick on the upper surface of the tongue, and thin on the lower. On the upper surface are a number of large papillæ; those on the posterior border, eight or nine in number, arranged like the letter V, are the largest, and called the *papillæ maximæ;* others are termed *capitatæ, mediæ, villosæ,* and *filiform;* the latter are the smallest, and are found principally at the middle of the tongue. The papillæ mediæ are the most abundant. The structure of all is the same. Their orifices are easily seen by the naked eye.

The following muscles compose the tongue: the *stylo-glossus,* the *hyo-glossus,* the *genio-hyo-glossus, lingualis, superficialis linguæ, transversalis linguæ,* and *verticales linguæ.* The first three form the principal part of its bulk.

Palate.—The palate is divided into two parts, the *hard*, and the *soft* palate. The first separates the mouth from the nose, and is composed of the palate processes of the upper maxillary and palate bones, covered by the common lining membrane of the mouth. It has transverse ridges extending to the alveolar processes, and also one in the median line.

The second, or *soft* palate, called also pendulous palate, consists of a loose membrane stretched transversely across the back of the mouth, at the posterior margin of the hard palate. Its middle portion, called the uvula, is free and projects downwards for a half or three-fourths of an inch; on each side of the uvula there are two crescenting folds of mucous membrane, called the *lateral half-arches*, the space between which is the *fauces*. Between the anterior and posterior arches on each side are the tonsil glands, which consist of a collection of mucous follicles. Each *tonsil* is about the size and shape of an almond.

There are several small muscles entering into the composition of the soft palate.

The *constrictor isthmi faucium*, a small muscle on each side, arising from the middle of the anterior half-arch, and inserted into the side of the root of the tongue. It diminishes the opening into the pharynx.

The *palato-pharyngeus*, is situated in the posterior half-arch, arising from its middle to be inserted into the side of the pharynx. It draws the soft palate downward and the pharynx upwards.

The *circumflexus*, or *tensor palati*, arises from the spinous process of the sphenoid bone and the Eustachian tube, and is inserted into the middle of the soft palate, and into the crescentic edge of the palate bone; its tendon winds round the hook of the internal pterygoid process. It extends or spreads out the palate.

The *levator palati* arises from the petrous portion of the temporal bone and the Eustachian tube, and is inserted into the soft palate. It draws the palate upwards.

The *azygos uvula* is in the centre of the soft palate and uvula. It shortens the uvula.

Salivary Glands of the Mouth.

There are three salivary glands situated on each side of the neck, bordering on the mouth, for the secretion of saliva. They are the parotid, submaxillary and sublingual. Their color is of a light pink.

The *parotid* is the largest, and is very irregular in shape. It occupies the space behind the ramus of the lower jaw, and the mastoid process. Externally it is covered by the skin; it is of a lobulated structure. Its duct, called the duct of Steno, is about the size of a crow's quill, and passes along the outer face of the masseter muscle, in a line drawn from the lobe of the ear to the end of the nose, and perforates the cheek by a very small orifice, opposite the second molar tooth of the upper jaw. The external carotid artery and the temporal vein pass up through its deeper portion; and the portia dura nerve also traverses it from behind forward.

The *submaxillary* gland is not more than half the size of the former, and is somewhat ovoidal in form, and lobulated in structure. It is situated in the depression on the inner face of the lower jaw, and covered externally by the skin, superficial fascia and platisma myoid muscle.

Its duct, called the duct of Wharton, empties into the mouth, under the tongue, at the anterior margin of the frænum linguæ, by a very small orifice.

The *sublingual* gland, which is smaller than the last, oblong in shape, lobulated in structure, is situated under the tongue, covered only by the lining membrane of the mouth. It discharges by several ducts near the duct of the submaxillary gland.

The Pharynx and Œsophagus.

The *pharynx* is a large membranous cavity situated between the vertebral column and the posterior part of the nose and mouth, and extending from the base of the cranium to the fourth or fifth cervical vertebræ. It is about five inches long, and funnel shaped, being larger above than below; its muscular attachments keep it always open.

It has three coats, a muscular, a cellular and a mucous one. The muscular coat is internal, and consists of three muscles on each side, placed one above the other, and called constrictors.

The *inferior constrictor* arises from the side of the cricoid and thyroid cartilages, and is inserted into its fellow of the opposite side on the back of the pharynx; the upper fibres are oblique, the lower horizontal.

The *middle constrictor* arises from the cornu of the os hyoides and the lateral thyreo-hyoid ligament, and is inserted into its fellow at the posterior median line.

The *superior constrictor* arises from the pterygoid process of the sphenoid bone, from the upper and lower jaw, from the buccinator muscles and the root of the tongue; and is inserted into its fellow behind, and also into the cuneiform process of the os occipitus. Its fibres are more horizontal than those of its fellows.

The successive contractions of these muscles convey the food from the mouth into the œsophagus.

The *stylo-pharyngeus muscles*, described among those of the neck, by contracting shortens the pharynx.

The *cellular* coat connects the muscular with the mucous, and merely serves for the transmission of nerves and blood-vessels.

The *internal* or *mucous* coat, which is a continuation of that of the mouth, nose and Eustachian tube, like it is covered by a delicate epidermis, and studded with mucous follicles and glands.

The *Œsophagus* is the canal situated between the spine and larynx, which conveys the food from the pharynx to the stomach. Its length is about nine or ten inches; and its diameter when inflated, near an inch, though this is not uniform, as it gradually increases as it descends. In descending, it passes through the posterior mediastinum, and at its lower part, where it passes through the diaphragm, inclines slightly to the left side. Its upper part is the narrowest portion of the alimentary canal; and consequently foreign bodies, if not arrested in the neck, readily pass through the remainder of the alimentary canal.

It is composed, like the pharynx, of three coats, a muscular, a cellular, and a mucous. The *muscular* is external, and thicker than any other portion of the canal; it consists of two layers, the fibres of the external layer being longitudinal; those of the internal, circular.

The *cellular* coat unites the others. The *mucous* coat is continuous with that of the pharynx, has a thick epithelium, contains numerous mucous glands and follicles, and when in a state of rest, presents a number of longitudinal folds.

Fig. 21.

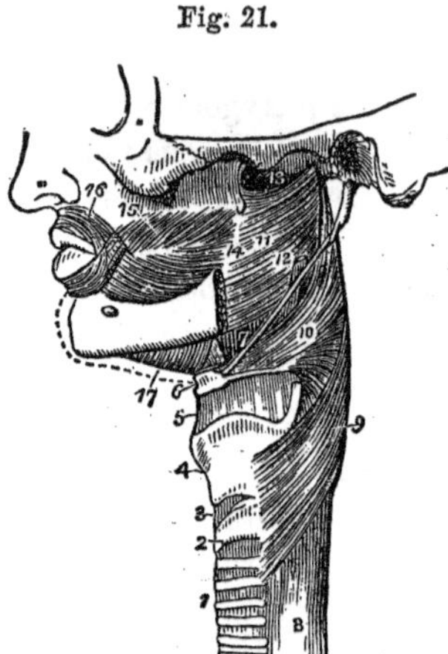

Deglutition, or swallowing, is effected by the contraction of the longitudinal muscular fibres, which shorten the passage, and by the successive contractions of the circular fibres from above downwards. In vomiting, the contractions commence below and go upwards.

Fig. 21. Lateral view of the pharynx. 1 Trachea; 2 cricoid cartilage; 3 crico-thyroid membrane; 4 thyroid cartilage; 5 thyro-hyoid membrane; 6 os hyoides; 7 stylo-hyoid ligament; 8 œsophagus; 9 inferior constrictor; 10 middle constrictor; 11 superior constrictor; 12 stylo-pharyngeus muscle; 13 upper concave margin of the superior constrictor; 14 pterygo-maxillary ligament; 15 buccinator muscle; 16 orbicularis oris; 17 mylo-hyoideus.

Of the Abdominal Viscera.

The cavity of the abdomen extends from the diaphragm above to the pelvis below; and is divided for descriptive purposes into nine different regions, by the drawing of two parallel vertical lines through the anterior inferior spinous processes of the ilia, and intersecting these by two other horizontal lines, the one drawn over the crests of the ilia, and the other over the most prominent part of the cartilages of the ribs. This makes three regions above, three in the middle, and three below. The region in the centre of the upper row is called the *epigastric*, and contains the left lobe of the liver and a portion of the stomach;

those on either side are termed the *right* and *left hypochondriac*, the first contains the right lobe of the liver, and the second contains the spleen, and a portion of the stomach and liver. The central region of the middle row is called the *umbilical*, and contains the small intestines; those on the sides are the *right* and *left lumbar*, the first contains the right kidney and ascending colon, the second the left kidney and descending colon. The region in the centre of the lower row, is the *hypogastric*, and contains a portion of the small intestines and the bladder; those on the sides are the right and left iliac regions, or the iliac fossæ, of which the first or right contains the *cæcum* or *caput coli*, and the second the sigmoid flexure of the colon.

Peritoneum.—The whole of the interior surface of the abdomen is lined, and the contained viscera covered by a thin, transparent, serous membrane, called the peritoneum, which, like all the serous membranes, is a closed sack. Its office is to secrete a small quantity of fluid to lubricate the viscera, and thus enable them to move readily upon each other, and the walls of the cavity; it also forms ligaments and connections by which the viscera are held in their places. Although the viscera appear to be contained within the cavity of the peritoneum, they are not so in reality, but are all on its outside. A familiar and good illustration of the manner in which the various viscera are covered by the peritoneum, is afforded in the application of a double night cap to the head. One part of the cap is close to the head and compares with the peritoneal coat of one of the viscera; the other is loose and compares with the peritoneum, where it comes in contact with the walls of the abdomen.

Those portions of the peritoneum, which pass between one viscus and another, generally consists of two lamina, and are called *omenta.*

There are four of these processes or omenta, viz. the *lesser omentum*, or gastro-hepatic, passing between the stomach and liver, and attached to the lesser curvature of the stomach; the *great omentum*, or gastro-colic, which extends between the stomach and colon, being attached to the greater curvature of the former, and including the latter between its lamina—it is the largest of the omenta, and is spread over the intestines like an apron — it is known as the *caul,* and in corpulent persons contains a great deal of fat; the *colic-omentum* or *mesocolon,* which holds the large intestine to the posterior wall of the abdomen; and the *gastro-splenic,* which extends from the stomach to the spleen.

The *mesentery* serves to connect the small intestines to the walls of the abdomen; it consists of two laminæ of peritoneum, and its inferior edge equals in length the entire intestine. The superior mesenteric arteries and veins, lymphatic glands and vessels, branches of the sympathetic nerve, fat, &c., are contained between the lamina of the mesentery.

Stomach.—The stomach is situated in the epigastric and left hypochondriac regions. It is a conoidal sac, having an upward curve; it has an anterior and posterior face, being somewhat flattened in front and behind; two curvatures, the upper of which is the lesser, and the lower the greater; two orifices, of which one, called the *cardiac,* is at the superior part of the left extremity, and is a continuation of the œsophagus into the stomach, and the other, the *pyloric,* is at the right extremity, and is continuous with the small intestine; and two extremities, of which the left is much the largest, and is a rounded cul-de-sac or tuberosity, and the left is a gradual diminution of the organ from its middle to the duodenum. Near the right end of the stomach is a dilatation, which is sometimes called the antrum pylori.

The stomach is retained in its place by the hepatico-gastric and the gastro-splenic omentum, and by its continuity with the œsophagus and duodenum. Its dimensions are variable, depending much upon the mode of life; generally, however, its capaciousness is between that of a pint and a quart. It is composed of four coats, a peritoneal, muscular, cellular, and mucous.

The *peritoneal coat* affords a complete investment for the stomach, and is closely attached to it, except at the curvatures where its looseness allows of the distention of the other coats.

The *muscular coat* is thicker than that of the intestines, but not so thik as that of the œsophagus; its fibres are arranged both in a longitudinal and circular direction. The longitudinal fibres are chiefly found along the lesser curvature, and the circular fibres, which cover the entire organ, are most numerous near the pyloric extremity.

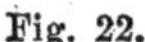
Fig. 22.

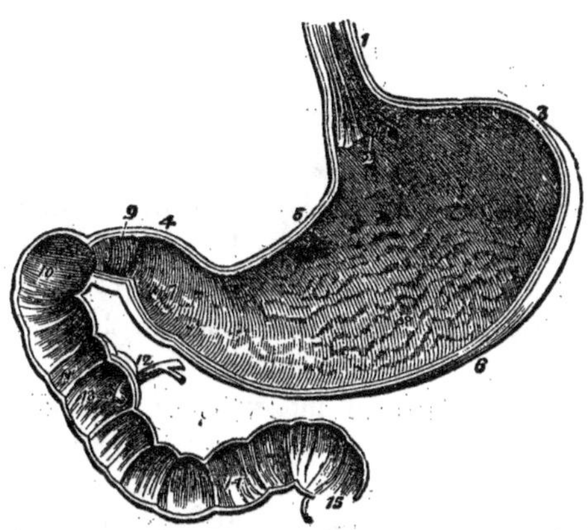

Fig. 22. A vertical and longitudinal section of the stomach and duodenum. 1 Œsophagus; 2 cardiac orifice of the stomach; 3 greater end of the stomach; 4 Lesser or pyloric end; 5 lesser curvature; 6 greater curvature; 7 antrum pylori; 8 folds or rugæ of the mucous membrane; 9 pyloris; 10 oblique portion of the duodenum; 11 descending portion; 12 pancreatic duct, and ductus communis choledochus; 13 opening of these ducts; 14 transverse portion of the duodenum; 15 commencement of the jejunum; the valvulæ conniventes are shown in the duodenum and jejunum.

The *cellular coat* connects the muscular with the mucous, and serves for the transmission of blood-vessels and nerves to the mucous coat.

The *mucous coat*, also called *villous*, is the most internal, and is continuous with that of the œsophagus; it is of a light pink color, and presents a velvety appearance, hence the name villous has been applied to it. When the stomach is not distended, this coat is thrown into longitudinal folds or *rugæ*, which are most numerous along the greater curvature, and near the pyloric orifice. A circular fold of mucous membrane at the pyloric orifice, which lessens its size, is called the *pyloric valve*. The office of this coat is to secrete the gastric juice.

The Intestines.

The whole length of the intestinal canal is from thirty to thirty-five feet. It extends from the pyloris to the anus, and is divided into large and small intestine.

Small Intestine.

The small intestine commences at the pyloris, and terminates by a lateral opening into the large intestine in the right iliac region. It comprises four-fifths of the length of the whole canal, and is from twenty to twenty-four feet long. It is cylindrical in shape, and about an inch in diameter, although there is a gradual decrease in diameter from above downwards.

Like the stomach, the small intestine has four distinct coats.

The *peritoneal coat* is the most external; and after completely investing the intestines, it is continued in two laminæ to be attached to the lumbar vertebræ, thereby constituting the mesentery.

The *muscular coat* is next to the peritoneal, and is thin and pale; the superficial fibres are longitudinal, though indistinct; the internal are circular.

The *cellular coat*, like that of the stomach, connects the peritoneal with the mucous, and serve to transport blood-vessels, nerves and lacteals. When dried it appears like cotton, as also does the corresponding coat in the other portions of the alimentary canal.

The *mucous coat* is internal, and much longer than the others, which allows it be thrown into numerous folds or duplicatures; these folds generally pass quite around the intestine, and overlap each other, like the shingles of a house. They are called *valvulæ conniventes*, and their office is to retard the progress downwards of alimentary matter, and to increase the absorbing and exhaling surface. The surface of the mucous membrane is covered with numerous papillary projections, called villi; and each villius is composed of an artery, vein and lymphatic. The lymphatics do not open directly upon the surface of the mucous membrane, but the chyle is conveyed out of the intestine by the intervention of cells.

The cellular coat is studded with mucous glands, whose ducts open upon the surface of the mucous coat. They are mostly microscopical. Some of them, which are visible to the naked eye, and called the *glands of Brunner*, are found scattered throughout the intestine, although existing in most abundance in the duodenum. Others still larger, called *Peyer's glands*, are found chiefly in the lower part of the small intestine. They consist of a cluster of smaller glands, and are consequently often called aggregated. Those of *Brunner* are also sometimes called *solitary*.

The small intestine is divided by anatomists, though without much reason, as it is a continuous tube, into the duodenum, jejunum, and ilium.

The *duodenum*, so named from its being about twelve fingers' breath, or twelve inches long, is that portion next the stomach. Its direction is curved, forming a segment of a circle, the concavity of which looks towards the left side. The mucous coat of the duodenum is tinged with bile, and contains a great number of valvulæ conniventes, and the glands of Brunner.

The *ductus communis choledochus*, from the liver and pancreas, empties into the duodenum, about four inches from the pylorus.

The *jejunum* and *ilium* form the remaining three-fifths of the small intestine, the former being two, and the latter one-fifth of the length; the only differences in their appearance are that the jejunum contains a large number of valvulæ conniventes, and is rather larger in its diameter than the ilium.

The small intestine, as stated, is attached to the posterior part of the abdomen by a process of peritoneum, called the *mesentery*. This attachment, called the root of the mesentery, is about six inches in length, and extends from the left side of the second lumbar vertebræ to the right iliac fossa.

Large Intestine.

The large intestine receives the effete matters from the small, and comprises about one-fifth of the length of the whole intestinal canal. It commences at the inferior end of the small intestine, and describing a circle, which surrounds two-thirds of the abdomen, terminates at the anus. Its diameter is much greater than that of the small intestine, and it also presents a sacculated appearance.

It is composed, like the small intestine, of four coats.

The *peritoneal coat* is continuous with the mesocolon and affords a complete investment, except at the lower part of the rectum, and the descending portions of the colon, where the latter comes in contact with the abdomen.

The *muscular coat* consists of longitudinal and circular fibres; the former are collected into three fasciculi, or bands, which extend to the rectum.

The *cellular coat* unites the muscular and mucous coats, and contains the blood-vessels and nerves.

The *mucous coat* is smooth, having neither villi nor valvulæ conniventes. It contains numerous follicles, called the follicles of *Lieberkühn*. It also contains some solitary glands.

The large intestine is divided into three parts, the cœcum, the colon, and the rectum.

The *cœcum* is the commencement of the large intestine, and is about two inches in length. It is confined by the mesocolon in the right iliac fossa. It is often called the *caput coli*, or head of the colon. Attached to its rounded extremity is a worm-like process of intestine, from three to four inches long, called the *appendix vermiformis*. This appendix is usually filled with flatus.

At the side of the cœcum is an elliptical opening, called the *ilio-colic valve*, by which the small intestine empties into the large. When the cœcum is distended, this valve becomes closed, and prevents the return of fœcal matter into the small intestine.

The *colon* comprises the principal part of the large intestine; it commences at the ilio-colic valve, and ascends on the right side of the abdomen to the margin of the false ribs, it then passes transversly across, beneath the stomach, to the left side, whence it descends to the left iliac fossa, and terminates in the *sigmoid flexure.*

The *rectum* commences at the left sacro-iliac symphisis, at the termination of the long, loose convolution of the colon, called the sigmoid flexure, and passes down in front of the sacrum to the anus.

Its *muscular coat* is much thicker and redder than that of any other portion of the intestines; the external fibres are longitudinal, and the internal circular. At the lower extremity of the rectum, the circular fibres are multiplied so much as to form a complete internal sphincter muscle.

The *mucous coat* is thick, red, and spongy, and five or six inches above the anus on each side, is thrown into a semi-circular fold, somewhat resembling the valvulæ conniventes, which in some degree prevent the involuntary discharge of fæces. Just above the anus are a number of small pouches, having their orifices pointing downwards. The rectum is larger in the middle than at the ends.

The Liver.

The liver occupies the whole of the right hypochondriac, and a portion of the epigastric and left hypochondriac regions. It is the largest glandular organ in the body, and secretes the bile. In shape it is oblong and oval; its weight is from four to five pounds, and it measures about ten inches in length, six or seven in width, and four or five in thickness. Its long diameter is across the body. Its color is a reddish brown, with occasional blue or black spots on its under surface and about its edges.

The upper surface is regularly convex, fitting closely to the concave under surface of the diaphragm; and the lower concave; the right end is also much thicker than the left. A broad ligament, called *suspensory,* formed from its peritoneal covering, holds the liver in contact with the diaphragm. The anterior part of this ligament is the ligamentum teres, and the posterior the coronary ligament. In the anterior edge of the liver is a notch, and in the posterior edge a deep depression for the spinal column. On its under surface, extending from the notch in front to the depression behind, is the *umbilical fissure*, or sulcus, so called from having accommodated the umbilical vein in the fœtal state; at right angles with this is the *transverse fissure*, which contains the hepatic artery and duct, surrounded by cellular membrane, and called the *capsule of Glisson*.

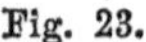

Fig. 23.

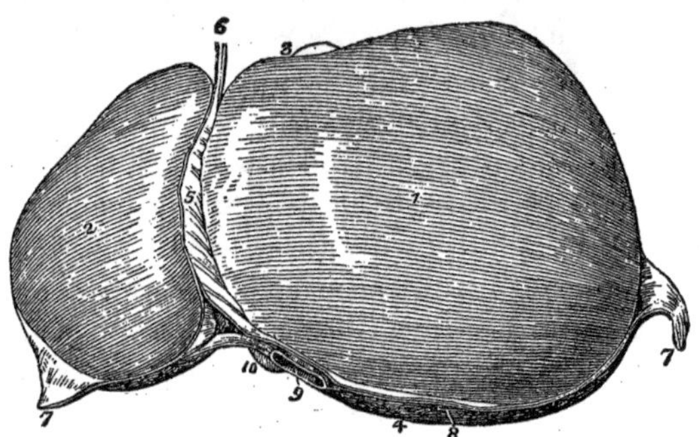

Fig. 23. Representation of the upper surface of the liver. 1 Right lobe; 2 left lobe; 3 anterior, free border, and fundus of gall bladder; 4 posterior border; 5 broad ligament; 6 round ligament; 7, 7 lateral ligament; 8 origin of the coronary ligament; 9 inferior vena cava; 10 point of the lobulus Spigelii.

The liver is divided into two lobes, a right and left lobe, the division being marked above by the suspensory ligament, and below by the umbilical fissure. The right lobe is much the larger, and has several elevations on its under surface; the principal of which are the *lobulus Spigelii*, *lobulus quadratus*, and the *gall bladder*.

Four sets of vessels ramify through the substance of the liver, making it extremely vascular.

The *portal vein* collects the blood from the stomach, intestines, pancreas, and spleen, and after reaching the transverse fissure, divides into two branches, called the right and left sinuses, one of which is distributed to each lobe of the liver.

The *hepatic artery,* also, conveys blood to the liver; it is a branch of the cœliac, and at the transverse fissure divided into three or more branches previous to penetrating the substance of the liver.

The *hepatic veins* arise by capillaries in the acini of the liver, and after collecting into three large trunks, empty into the ascending vena cava at the posterior margin of the liver. These veins are destitute of valves, and have very thin parietes.

The *hepatic duct* also commences by capillaries in the liver; when it reaches the transverse fissure it is about the size of a writing quill. It joins the duct of the gall bladder at an acute angle; and the union of the two forms the ductus communis choledochus, which empties into the duodenum three or four inches from the stomach.

Fig. 24.

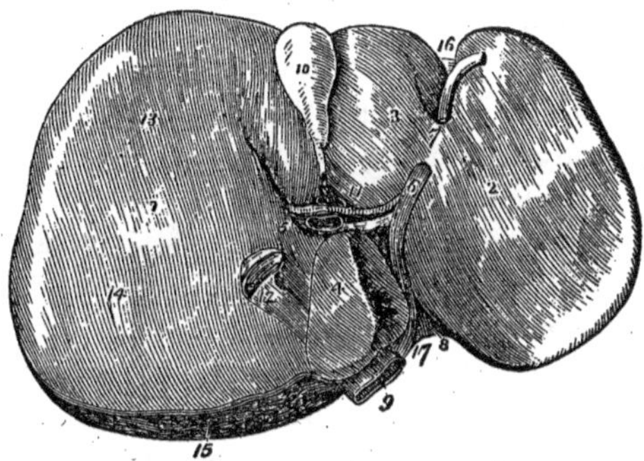

Fig. 24. Representation of the under surface of the liver. 1 Right lobe; 2 left lobe; 3 lobulus quadratus; 4 lobulus Spigelii; 5 lobulus caudatus; 6 longitudinal fissure; 7 pons hepaticus; 8 fissure for the ductus venosus; 9 inferior vena cava; 10 gall bladder; 11 transverse fissure; 12 vena cava; 13 depression for the curve of the colon; 14 double depression made by the right kidney and its supra-renal capsule; 15 rough surface on the posterior border of the liver; 16 notch separating the lobes anteriorly; 17 depression on the posterior border for the spinal column.

The minute *structure* of the liver is made up of a numerous collection of irregular lobules, or *acini*. This arrangement is best seen on tearing the organ. Each lobule is said to be about the size of a millet seed, and to represent of itself a perfect gland, being formed by the termination of the blood-vessels, and by the origin of a branch of the hepatic duct, called the porus biliaris. These lobules are held together by means of a cellular tissue, which is called the parenchyma.

The Gall Bladder.

The gall bladder is the reservoir for the bile; it is attached to the under surface of the right lobe of the liver to the right of the umbilical fissure, and has its long diameter inclining slightly to the right side. It is a pyriform sac, and about three inches in length. Its rounded extremity, called the *fundus*, projects somewhat beyond the anterior border of the liver. Its posterior extremity, or *neck*, is narrow and twisted to retard the passage of a fluid through it.

It has three coats, a peritoneal, a cellular, and a mucous one. The peritoneal coat is but a partial one, covering its inferior surface only.

The middle is of strong cellular membrane; and the internal mucous coat is thrown into irregular delicate folds, and tinged of a deep green or yellow color by the bile.

The *duct* of the gall bladder, called the *cystic*, is shorter and smaller than the hepatic, which it joins at an acute angle; the union of the two forming the ductus communis choledochus, which is from two to three inches long, and about the size of a goose quill. As mentioned it empties into the duodenum, in an oblique manner, after passing through the right extremity of the pancreas. Its orifice is very small, and marked by a tubercle on the inner side of the duodenum.

The *bile* is of a deep yellow, sometimes green color, and bitter taste. When recently secreted, it is thin and fluid; but after remaining for some time in the gall bladder it becomes as thick as molasses, and increases, also, in the intensity of its color and in bitterness. This is owing to the absorption of its watery particles by its mucous coat.

The chief use of the bile is to aid digestion by dissolving the fatty matters, and rendering them capable of being taken up by the lacteals.

The Spleen.

The spleen is situated in the posterior part of the left hypochondriac region. Immediately above it is the diaphragm, below it the colon, and to the right the large end of the stomach and the pancreas. It is of a semi-oval figure, and of a deep blue or brown color. Its external surface is convex; its internal slightly concave, and has an imperfect fissure in its centre, where the blood-vessels enter. Sometimes, also, its margins are notched. The usual size of the spleen is four or five inches in length, by two or three in breadth. In some individuals several spleens are found, the additional ones in such cases being quite small. It is kept in its place by ligaments, formed of the peritoneum, which pass between it and the diaphragm, stomach and colon.

In *structure* it presents a dark, brown pulp, which is held together by cells formed of the internal coat. It has two coats, an external peritoneal one, and an internal thin, gray and elastic one.

The *splenic* artery, which is the largest branch of the cœliac, furnishes it with blood; its vein empties into the vena portarum.

The spleen has no secretion, and its use is not well ascertained.

The Pancreas.

The pancreas secretes saliva, and is the largest salivary gland in the body. It is placed horizontally across the spine, in front of the last dorsal and first lumbar vertebræ, and behind the stomach. Consequently it is in the lower back part of the epigastric region. Its length is from six to seven inches, and its width about two. It is flattened before and behind. The anterior face looks obliquely upwards, and the posterior face obliquely downwards. The right extremity is enlarged into a head or tuber, sometimes called the lesser pancreas, and is in contact with the curvature of the duodenum; the left extremity is connected with the spleen.

The color of the pancreas, like that of the other salivary glands, is of a light gray or pink hue; and its structure is lobulated, the lobules being held together by intermediate cellular tissue.

It has no peritoneal coat, but is included between two lamina of the mesocolon. Its duct, called sometimes the duct of *Wursungius,* empties into the duodenum near the orifice of the ductus communis choledochus, and sometimes into the latter. The arteries of the pancreas are branches of the splenic, and its veins also empty into the splenic.

Physiology of Digestion.

Before the food is fitted for purposes of nutrition, it has to undergo several changes in the digestive organs.

The first stage in the process of digestion is called *prehension,* or the taking of food into the mouth. This is chiefly accomplished by the hand, some assistance being rendered by the front teeth, lips, cheeks and tongue.

The second stage is *mastication* (chewing), by which the food is rendered sufficiently fine to be taken into the stomach, and acted on by its juices. The food is kept between the teeth by the lips and cheeks externally, and by the tongue internally, and the closure of the soft palate upon the tongue prevents it from passing into the œsophagus. The motion of the lower teeth upon the upper reduce it to the requisite fineness.

While the food is being masticated, it is mixed with the saliva and juices of the mouth, which soften it and aid in its reduction. This is called *insalivation.*

The third stage is *deglutition* or swallowing, which, when the food has been sufficiently comminuted and moistened in the mouth, is effected by a simultaneous action of the muscles of the tongue, cheeks, floor of the mouth, soft palate and pharynx. The elevation of the soft palate prevents the food from passing into the posterior opening of the nostrils (nares); and it is prevented from passing from the pharynx into the trachea by the closure of the glottis (opening of the trachea) by the epiglottis. After entering the œsophagus, the alternate relaxing and contraction of the circular bands of muscular fibres gradually force it downwards into the stomach. In the effort of vomiting, the action of the œsophagus is reversed—the contraction of its muscular fibres commencing at its lower extremity. The spasmodic cough which sometimes takes place while eating or drinking, is owing to some particles of the aliment entering the larynx or trachea, which cannot occur unless the glottis is opened by the inhalation of air. An attempt to speak when the mouth contains food is generally the cause.

The fourth stage in the process of digestion is *chymification*, or the conversion of the food into a homogeneous pulpy mass, generally of the consistence of cream, called *chyme*. This change is effected in the stomach; as soon as the food enters this organ, it is thoroughly mixed with the gastric juice, by the alternate contraction and relaxation of the fibres of the muscular coat, which produce a great variety of motion. The contraction of the muscular coat as well as the secretion of the gastric juice from the follicles of the mucous coat, is occasioned by the stimulus of the food. No gastric juice is contained in the organ when in a state of rest, but its secretion is always excited by the presence of a foreign body. While the process is going on, the food is prevented from escaping from the stomach by the closure of its orifices.

The gastric juice is colorless, slightly viscid, and has an acid reaction. By analysis it has been found to contain free muriatic and acetic acids, and phosphates and muriates of potassa, soda, magnesia and lime. Although the presence of dilute acids in the gastric fluid is essential to its action, the active agent is an organic compound, called *pepsin*.

This agent undergoes no change itself, but induces changes in other substances, disposing them to dissolve in the acids of the stomach, and form with them definite chemical compounds.

The degree of solubility of different substances in the juices of the stomach varies. As a general rule, animal food is more soluble than vegetable—though this is not invariably the case. Fatty substances undergo but little change in the stomach; they require admixture with the bile, whereby a saponaceous compound is formed, to fit them for absorption by the chyliferous vessels.

The time required by the stomach to convert the food into chyme, varies from three to five hours according to the nature of the food.

The fifth stage is *chylification.* After the food is reduced to chyme in the stomach, it passes through the pyloric orifice into the duodenum, where it is mixed with the bile and pancreatic juice, and by their action converted into chyle and residual matters. The chyle is a whitish or whey-like fluid, with a creamy pellicle. As the contents of the intestine are gradually carried along the canal by an action, called *peristaltic*—induced by the alternate contraction and relaxation of the muscular coat—the chyle is absorbed by the lacteals through the villi of the mucous coat, and the residual fœcal matters pass into the large intestine, where it accumulates until finally expelled from the system. After leaving the intestinal canal, the chyle is carried by the lacteals through the mesenteric glands or ganglia, and emptied into the receptaculum chyli at the commencement of the thoracic duct, along which it passes to enter the circulation at the junction of the internal jugular and subclavian veins.

The Kidneys.

The kidneys are two glands, for the secretion of urine, one of which is situated on each side of the spine, in the back part of the lumbar region. They extend from the last dorsal to the third lumbar vertebræ; the right one being somewhat lower than the left, to accommodate the right lobe of the liver. In shape the kidney is oval, resembling the kidney bean; its position is upright, with the excavation, called *hilum,* presenting towards the spine. It is a hard, solid body of a reddish brown color. The length is about four inches, and the breadth two.

The kidneys are placed without the cavity of the peritoneum, and surrounded with an abundance of fat, and cellular tissue. Their proper covering is a dense fibrous capsule which envelops them completely, and penetrates into the fissures.

The kidney consists of two different *structures*, which may be readily seen by cutting it open longitudinally. One, the external, is called *cortical*, the other, the internal, *medullary*. The *cortical substance* is about one-fourth of an inch thick, and forms the whole circumference of the kidney. It consists of a number of tortuous tubes of *Ferrein*, which secrete the urine, between which are numerous minute blood-vessels.

The *medullary* substance is of a darker color than the cortical, and consists of from twelve to eighteen cones (named after *Malpighi*), which are arranged in three rows with their apices converging towards the *hilum* or fissure. These apices are called papillæ renales, in consequence of their projecting like so many small nipples. Each cone forms a sort of distinct gland, and can be subdivided into numerous tubes, called the tubes of *Bellini*, into which the tortuous tubes of *Ferrein*, from the cortical substance, empty. As the urine oozes from the orifices of the papillæ, it is received into a membranous cup, the *infundibulum*, which surrounds each papilla. From these infundibulæ, four or five of which are united into a common trunk for that purpose, and called a *calyx*, the urine passes into a common receptacle, called the *pelvis*, which is formed by the junction of about three calyces.

The *pelvis* is in the centre of the kidney, and of a conoidal shape; from it the excretory duct of the kidney, called the ureter, conveys the urine to the bladder. The ureter is cylindrical, and about the size of a writing quill; it has two coats, an internal mucous one, and an external fibrous one. They are thin, white, and extensible.

The *ureter* descends between the peritoneum and psoas magnus muscle, and enters the inferior fundus of the bladder obliquely, by a very small orifice.

Supra-Renal Capsules.

There are two small bodies, placed one upon the upper end of each kidney. They are of a yellowish brown color, and triangular pyramidal shape, the base being concave to rest upon the kidney. They have no excreting duct, nor is their use known. It is supposed, as they are larger in the fœtus than the adult, that their use is confined to fœtal life.

The Bladder.

The bladder receives and serves as a reservoir for the urine. It is placed in the pelvis immediately behind the pubes. When distended, its shape is generally oval, the large end being downwards; in women and young children it is nearly spherical; it is also, mostly, more capacious in woman than in man. The dimensions of the bladder varies; when healthy it will generally hold near a pint. It is bounded in front by the pubes, above by the small intestine, behind by the rectum, and below by the prostate gland and seminal vesicles. A conical ligament, called the *urachus*, extends from the superior extremity of the bladder to the umbilicus; on each side of the urachus, in the folds of the peritoneum, are the round ligaments, which were the umbilical arteries in fœtal life.

Besides these ligaments, the pelvic aponeurosis also assists to retain the bladder in its place. Its upper extremity is called the *superior fundus*, its lower extremity the *inferior fundus*, and between the two is the body. The neck is at the junction of the bladder with the urethra.

The walls of the bladder consists of four coats, viz. a peritoneal, a muscular, a cellular, and a mucous one.

The peritoneal coat covers the superior fundus, and the posterior part of the body. The muscular coat consists of pale fibres which pass in various directions. It is somewhat thicker than that of the intestines; and its contractions expel the urine.

The cellular coat connects the muscular and mucous coats, and is dense, strong, capable of much distention, and impervious to water. Through it the vessels and nerves are transmitted.

The mucous coat, also called the villous, is much more smooth than that of the intestines; it is of a light pink color, and contains a great number of small mucous follicles. An angular space within the cavity of the bladder, included between the orifices of the ureters and the orifice of the urethra, is called the *vesical triangle.* A projection in the anterior angle, caused by the third lobe of the prostate, is called *uvula vesicæ.*

The *sphincter vesicæ* consists of a semi-circular and transverse band of muscular fibres surrounding the neck of the bladder. Its office is to keep the orifice closed.

The *urethra* is the membranous canal which conducts the urine from the bladder, in the male it also conducts the semen; it consists of an internal mucous coat, continuous with that of the bladder, and an external muscular one. Its course is curved. In the male, the portion which extends from the neck of the bladder through the prostate gland, is called the *prostatic* portion, and is about an inch in length; in it are the uvula vesica, and the caput gallinaginis or verumontanum—the latter is a triangular elevation of mucous membrane, at the base of which are the orifices of the ejaculatory ducts. The next portion is about three-fourths of an inch in length, quite narrow, and is the membranous portion; the remaining portion, which is the longest, passes through the penis; when it enters this organ, there is an enlargement, called bulbous. The *urethra* of the female is much shorter than that of the male; it passes from the neck of the bladder downwards and forwards under the symphysis pubis, and has its sexternal orifice at the superior, anterior angle of the vagina. The orifice is marked by a slight elevation.

The Prostate Gland.

This is a hard body, about the size and shape of a horse chestnut, fixed to the neck of the bladder. It rests upon the rectum behind, and in front is bounded by the triangular ligament. It consists of three lobes, through the middle one of which the urethra passes. It secretes a thick, white mucous, which is discharged into the urethra.

Organs of Respiration.

These are the larynx, the trachea, and the lungs.

Larynx.—The larynx is a cartilaginous tube, forming the commencement of the windpipe. It is placed in the upper and anterior part of the neck immediately below the os hyoides. The œsophagus is situated immediately behind it, separating it from the vertebræ of the neck; and on each side are placed the primitive carotid arteries and the internal jugular veins. It gives passage to the air to and from the lungs, and also regulates the voice; its superior portion is prismatic, and its inferior circular. In males it is larger, proportionably, than in females.

The five following cartilages compose it, viz. one thyroid, one cricoid, two arytenoid, and one epiglottis.

The *thyroid* is the largest of the five, and is placed at the upper and anterior part of the neck, about one inch below the os hyoides. It consists of two symmetrical, quadrilateral plates which unite in an acute angle at the median line, and constitute the prominence in the upper part of the throat known as the *pomum Adami* (Adam's apple), which is much larger in men than women. In the upper part of this prominence is a deep notch. On either side the superior margin is curved like the letter S, and to it is attached the middle thyreo-hyoid ligament; the inferior margin is also curved, but in a less degree, and to it is attached the middle crico-thyroid ligament. The posterior margin is elongated above and below into processes or cornua, of which the superior are the longer, and called cornua magna, and are attached to the lateral thyreo-hyoid ligament; and the inferior, which are short and curved, are called the cornua minor, and articulate with the lateral crico-thyroid ligaments.

The *cricoid* cartilage is situated below the thyroid, and forms the base of the larynx; it is an oval ring, with the lower margin nearly straight and horizontal, and connected to the first ring of the trachea; the superior margin is oblique on account of the breadth being three times as great behind as in front; on each side of the superior part of the posterior margin is a small convexity for articulating with the arytenoid cartilages. Externally the posterior surface is flattened, and from it arises the crico-arytenoid muscle.

The *arytenoid* cartilages are placed at the upper and posterior portion of the larynx, and in shape resemble triangular pyramids curved backwards. Their bases articulate with the cricoid cartilage. When joined together, the two cartilages resemble the mouth of a pitcher, from which they received their name; to the anterior surface, which is uneven, the superior and inferior thyreo-arytenoid ligaments.

The *epiglottis* cartilage is situated on the posterior face of the base of the hyoid bone, and is partially enclosed by the two sides of the thyroid cartilage. Its form is that of an oval disk; the upper edge is thin and rounded; the lower part is also thin and elongated, and attached to the thyroid cartilage. Its attitude is vertical, immediately behind the base of the tongue, and projecting somewhat above it; the anterior surface is slightly concave, and the posterior convex.

Besides these cartilages there is sometimes a small one, called corniculum laryngis, attached to the apices of each of the arytenoid cartilages.

Numerous ligaments hold these cartilages together. Four of these within the larynx, are the *thyreo-arytenoid ligaments;* the two inferior are commonly called the vocal cords, and extend one on each side from the angle of the thyroid to the base of the arytenoid cartilages; the space included between them is the *rima glottidis.* The two *superior* ligaments, also one on each side, are placed three or four lines above the inferior, and extend from the angle of thyroid to the middle of the arytenoid cartilages. All of these ligaments are small, round, fibrous threads, covered by a reflection of the lining membrane of the larynx. The action of the small muscles of the larynx render them more or less tense.

The following pairs of muscles belong to the larynx, viz. the *thyreo-hyoid,* the *cryco-thyroid,* the *posterior crico-arytenoid,* the *lateral crico-arytenoid*, the *thyreo-arytenoid*, the *oblique arytenoid*, the *transverse arytenoid*, the *thyreo-epiglottideus*, and the *aryteno-epiglottideus*.

Their names sufficiently indicate their attachement; they serve to move the various cartilages, and modulate the voice.

The larynx is lined internally by a mucous membrane continuous above with that of the pharynx, and below with that of the trachea.

Fig. 25.

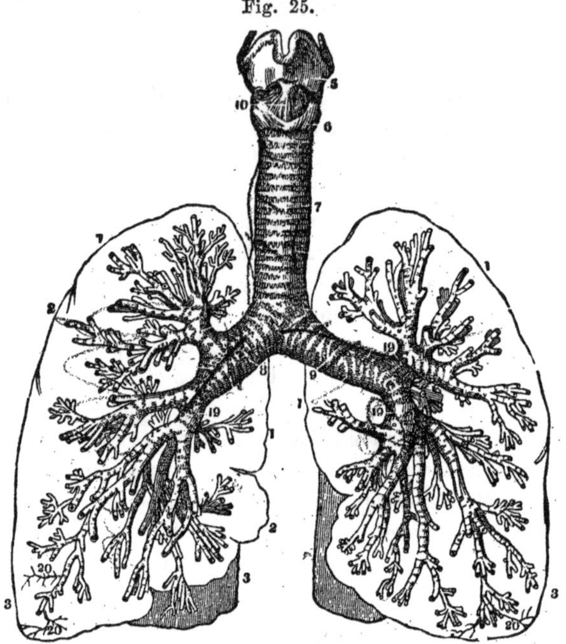

Fig. 25. A view of the larynx, trachea, and bronchia, with an outline of the lungs. 1, 2, 3 Outline of the lungs; 5 thyroid cartilage; 6 cricoid cartilage; 7 trachea; 8 right bronchus; 9 left bronchus; 10 muscles connecting the thyroid and cricoid cartilages; 19, 19 the subdivisions of the right and left bronchial tubes; 20, 20 capillary vessels.

The Trachea.

The trachea is a cylindrical canal, four or five inches in length by about three-fourths of an inch in diameter, extending from the larynx as low as the third dorsal vertebræ, and terminating in two ramifications called bronchia. It is composed of from sixteen to twenty distinct rings of cartilage united by an elastic ligamentous tissue. These rings are deficient in the posterior third of their circumference, which is completed by a muscular structure, whose fibres are transverse, and the contractions of which, diminish the diameter of the larynx, and thereby facilitate expectoration. Each ring is about a-fourth of an inch broad, and half a line thick. The trachea is lined by a mucous membrane continuous with that of the larynx.

The structure and arrangement of the bronchia is the same as that of the trachea. The right *bronchus* is larger and of a larger diameter than the left. At the orifice of each branch there is a semi-lunar cartilage, which forms somewhat more than half of its circumference, and the office of which is to keep the orifice open. The bronchia, after ramifying into numerous subdivisions, terminate in the lobules of the lungs. The structure of the smaller ramifications is somewhat modified; the cartilages, instead of being formed of one piece, are composed of several pieces, and are placed farther apart; finally, they disappear entirely, and the bronchia are membranous only.

The Thyroid Gland.

This gland is situated in front of the first and second rings of the trachea, and at the sides of the larynx. It consists of two lobes, one on each side, united by a thin, narrow portion stretched across in front of the upper part of the larynx, called the isthmus. It is of a dark brown color, and granular structure. It is also covered by a capsule which gives it a polish. Its use is not known.

In some districts of country it becomes very much enlarged, constituting the disease known as bronchocele or goitre.

The Lungs.

The lungs are the essential organs of respiration, and are situated one on each side of the chest, occupying the principal part of its cavity. They are conical in shape, and separated from each other by the heart and by a membranous septum, called the mediastinum; the diaphragm separates them from the abdomen. On the external surface they are concave, corresponding with the walls of the chest; internally they are concave to receive the convexity of the heart. The superior extremity is a tapering cone, terminating above the level of the first rib; the inferior extremity is broad and concave, and rests upon the diaphragm. The color of the lungs is a light pink, with specks or patches of black.

The right lung is somewhat shorter than the left, though more voluminous; and is divided into three lobes, while the left has but two.

Each lung is supported in its place by its root, which is composed of the pulmonary artery, pulmonary veins, bronchial tubes and vessels, and the pulmonary plexus of nerves.

In *structure* the lungs are divided into numerous lobules, and these again into minute air-cells, all of which are held together by cellular tissue, through which the blood-vessels, nerves and lymphatics ramify. The shape of the air-cells is polyhedral, and a certain number of them communicate with each other, by lateral openings, and with a single branch of the bronchial tube. They are lined by mucous membranes. The intermediate cellular substance is called parenchyma.

The pulmonary artery, which conveys the dark, venous blood to the lungs, terminates in capillaries which ramify on the walls of the air-cells; from these walls arise the pulmonary veins, by which the arterial blood, purified in its passage through the capilaries, is returned to the heart.

The bronchial arteries are the nutritious arteries of the lungs; they arise from the thoracic aorta.

The lymphatics, which are numerous, arising from the surface and the substance of the lungs, terminate in the bronchial glands, which are placed at the roots of the lungs, and at the bifurcation of the bronchia.

The nerves are branches of the sympathetic and eighth pair.

Fig. 26.

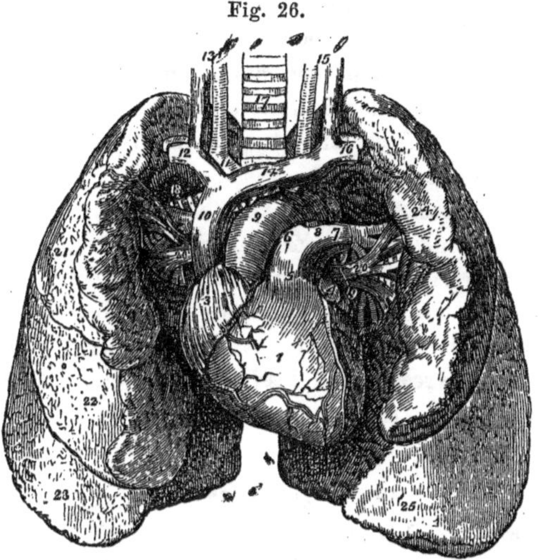

Fig. 26. Front view of the heart and lungs. 1 Right ventricle of the heart; 2 left ventricle; 3 right auricle; 4 left auricle; 5 pulmonary artery; 6 right pulmonary artery; 7 left pulmonary artery; 8 remains of the ductus arteriosus of the fœtus; 9 arch of the aorta; 10 superior vena cava; 11 arteria innominata; 12 right subclavian vein; 13 right common carotid artery and vein; 14 left vena innominata; 15 left carotid artery and vein; 16 left subclavian vein and artery; 17 trachea; 18 right bronchus; 19 left bronchus; 20, 20 pulmonary vein; 7, 19, 20 form the root of the left lung; and 18, 20 the root of the right lung; 21 superior lobe of the right lung; 22 its middle lobe; 23 its inferior lobe; 24 superior lobe of the left lung; 25 its inferior lobe.

The Pleuræ.

Each lung is invested by a serous membrane which maintains its structure and gives to it its shining appearance. After enclosing the lung as far as the root, it is reflected upon the parietes of the chest. The portion that covers the lungs is called the *pleura pulmonalis,* and that in contact with the chest the *pleura costalis.* The diaphragm and the root of the lung is also covered by it.

The Mediastinum.

The space included between the pleura of the two lungs is called the *mediastinum.* It is in the middle line of the thorax, and is divided into the anterior, middle and posterior.

The anterior mediastinum is the triangular space between the sternum and heart; it contains the remains of the thymus gland, and some loose cellular tissue.

The middle mediastinum contains the heart, the ascending aorta, the superior vena cava, the bifurcation of the trachea, the pulmonary arteries and veins, and the phrenic nerves.

The posterior mediastinum extends from the heart to the spinal column, and contains the descending aorta, the superior intercostal, and greater and lesser azygos veins, the thoracic duct, the œsophagus, the great splanchnic nerves, and the pneumogastric nerves.

The Thymus Gland.

The thymus gland is placed in the anterior mediastinum; it is of a triangular shape and pink color. Its structure is lobulated. Its use is not known; up to the second year of age it grows, after which it gradually diminishes and almost entirely disappears. In all probability it is of importance in fœtal life.

Physiology of Respiration.

The function of respiration consists in the conversion of venous into arterial blood. This change, by which carbonic acid is given off from, and oxygen taken into the system, takes place in the lungs. In the periphery of the body the circulating fluid becomes dark colored, and is taken up by the radicles of the veins, and after being mixed with the lymph and chyle, is carried to the lungs, where it is brought into contact with the atmospheric air, which restores its red color, and fits it for purposes of nutrition.

The manner in which this is accomplished is the following: the walls or sides of the air cells are very thin and transparent, and the capillaries are so placed between two adjacent cells, as to come in contact with the air from both. The carbonic acid of the venous blood then passes out through the walls of the air cells by *exosmose* or exudation, to unite with the nitrogen of the air, for which it has a strong affinity, and be exhaled from the lungs; and the oxygen of the air contained in the air cells passes through their parietes by *endosmose*, or imbibition, to unite with the blood, giving to it its bright red arterial color, thus vitalizing it, or fitting it for the various purposes of life.

The carbonic acid given off in respiration is chiefly furnished by the continual decay of the tissues. The carbon of the food, however, is also directly converted into carbonic acid to a considerable degree,—varying in quantity according to the amount of animal heat required.

Inspiration, or the action by which air is taken into the lungs, and expiration, or the action by which the air received in inspiration is expelled from the lungs, are performed by the expansion and contraction of the chest.

In the first inspiration, the diaphragm is the principal agent in the dilatation of the chest; when in a state of rest, this muscle is much arched, but by contracting it becomes more plane, and by flattening its arch increases the cavity of the thorax, while at the same time, by forcing down the abdominal viscera, it causes the protrusion of the abdomen, witnessed in inspiration. The intercostal and other of the muscles of the chest produce the lateral dilatation, which is however slight in ordinary natural inspiration, but is considerable when the inspirations are deep and forced. In a natural, quiet inspiration, the expansion of the chest is almost wholly accomplished by the diaphragm. This is also the case in old persons, the ossification of the cartilages of the ribs preventing lateral expansion.

In general there are from sixteen to eighteen inspirations per minute, though the number varies greatly under different circumstances. The average proportion, numerically, of the pulsations of the heart to the respiratory movements is about four and a half to one. At about every fifth inspiration the movements of the thorax are considerably increased.

In the second, or *expiration,* the parts concerned in inspiration return again to their natural state; the elasticity of the cartilages of the ribs, the rings of the bronchia, and of the air cells themselves, all aid in accomplishing this, as does also the contraction of the abdominal, and some of the thoracic muscles.

After each expiration, a short interval of repose succeeds; the length of this interval has been estimated as follows. By representing the whole period of time occupied in one respiratory act, from the beginning of one inspiration to the beginning of the next, by ten, the inspiration may by estimated at five, the expiration at four, and the interval of repose at one.

The amount of air inhaled at each inspiration is estimated to be about twenty cubic inches, which, allowing sixteen inspirations per minute, would give 19,200 cubic inches in one hour, passing through the lungs of an individual, or 460,800 cubic inches in twenty-four hours.

A certain portion of air, called *residuary,* always remains in the lungs, and upon it their lightness depends. It has been calculated that after an ordinary expiration more than a hundred cubic inches remain, and after the strongest expiration more than thirty. After the lungs have been once inflated by a full inspiration, no power whatever can remove the air from them so as to cause them to sink in water. The residuary air will not support life a longer time than three or four minutes.

"The movements of respiration are partly voluntary, and partly involuntary. Partly voluntary, in order that they may be inservient to the production of vocal sounds, and to the actions of speaking, singing, &c. Partly involuntary, lest in sleep or in moments of forgetfulness, the movements of respiration should be suspended, and fatal results ensue."

The nerves which govern the respiratory movements are derived from the *medulla oblongata* at the base of the brain.

Circulatory System.

The organs concerned in the circulation, are the heart, the arteries, the veins, and the lymphatics.

The Heart.

The heart, the centre of the circulation, is a hollow muscular organ, situated in the thorax between the sternum and the spine; being bounded at the sides by the lungs, and resting on the centre of the diaphragm below. In shape it is conoidal, having the apex inclined to the left side, and in contact with the walls of the thorax about the junction of the fifth rib with its cartilage. The side which rests upon the diaphragm is somewhat flattened. The length of the heart is about five inches and a half, and its diameter at the base three inches and a half. Its ordinary weight is about six ounces. It has four cavities, two of which are called *auricles,* and two *ventricles.* The heart has two functions, viz. to receive the blood and throw it into the lungs, and to receive it again after it has been *oxygenated* in the lungs, and distribute it throughout the body. The auricles are the receptacles of the blood, and the ventricles propel it through the system.

The auricles form the base, and the ventricles the body of the organ; the anterior extremity of the left ventricle, which extends somewhat beyond the right, constitutes its apex.

A membrane, called the *pericardium,* surrounds the heart, and also invests the roots of the large vessels connected with it. This membrane consists of two layers, an internal and an external one; and is only attached to the base of the heart, the remainder of the organ being only loosely enveloped by it.

The *right auricle* is an oblong cuboidal cavity, having superiorly an elongated process, which bears some resemblance to the ear of an animal, from which the term *auricle* has probably originated; anteriorly it has a convexity or pouch, called its *sinus.* Posteriorly, at its superior angle, the descending vena cava enters, and at its inferior angle the ascending vena cava. Its parietes are thin, and composed of muscular fibres arranged in parallel lines, resembling the teeth of a comb, whence their name *musculi pectinati*. Between the orifices of the two vena cava, an elevation exists called the *tuberculum Loweri*. A depression on the partition between the two auricles is called the *fossa ovalis;* in fœtal life an opening existed at this place, called the foramen ovale. Between the right auricle and ventricle is a round hole about an inch in diameter, called the *ostium venosum*, for the passage of the blood.

The *right ventricle* receives the blood from the right auricle. It is a somewhat triangular cavity, having its base downwards, with thick parietes, and is also larger than any of the other cavities of the heart. It is placed anterior to the left ventricle, from which it is separated by a thick septum. The internal surface of the cavity is composed of large fleshy fibres, called *columnæ carnæ*, from a number of which several tendinous chords (*chordæ tendineæ*) proceed to be inserted into the loose edge of the tricuspid valve. This valve is situated between the auricle and ventricle, opening into the latter, and is formed of a doubling of the lining membrane of the ventricle. It is circular, and attached around the ostium venosum; at its loose margin are three points or processes, from which it derives its name *tricuspid*. When the heart contracts, this valve closes the ostium venosum, and prevents the blood from returning into the auricle, and hence it passes into the pulmonary artery. The orifice of the pulmonary artery is circular, and about an inch in diameter; it is furnished with three valves, called from their shape *semilunar;* they are formed from the internal coat of the artery, and open outwards; in the centre of their loose edges is a small cartilage, called the *corpus aurantii*, the office of which is to perfect the closure of the valves. The use of the valves is to prevent the blood returning from the artery into the ventricle, when the latter dilates. Behind each valve is a pouch, called the *sinus of Valsalva*. The pulmonary artery passes upwards, and backwards to the under side of the arch of the aorta, and there divides into two branches, one for each lung, the branch for the right lung being the longest and largest. The diameter of the pulmonary artery is the same as that of the aorta, but its parietes are thinner.

The *left auricle* is concealed by the right and by the ventricles. Its shape is quadrangular or square, and into each of its four angles one of the pulmonary veins enter. Its parietes are muscular, smooth, and rather thicker than those of the right auricle. Its ear-like appendage is narrower and more crooked than that of the right auricle, and the *musculi pectinati* also enter into its structure. The opening between it and the left ventricle is likewise called *ostium venosum*. The partition between the auricles is sometimes imperfect in the adult.

The *left ventricle* is a conical cavity, and forms the apex of the heart. Its parietes are much thicker than those of the right ventricle. Internally, its surface is roughened by numerous *columnæ carnæ*, which become tendinous (*chordæ tendinæ*) where they are attached to the bicuspid or *mitral valve*. This valve, which consists of two folds or leaflets of the lining membrane of the ventricle, has its base attached to the margin of the ostium venosum, and its edges opening downwards into the ventricle: consequently when the ventricle contracts, it closes the opening, and the blood passes out by the aorta. The orifice of the aorta is supplied with three semilunar valves, arranged precisely like those at the mouth of the pulmonary artery.

The nutritious vessels of the heart are the right and left coronary arteries. The veins accompanying them empty into the right auricle. The nerves are branches of the sympathetic, supplied by the cardiac plexuses.

The *pericardium* is a membranous sac, consisting of two layers, in which the heart is loosely enveloped. The external layer is fibrous, white, and inelastic; the internal layer is serous, lines the external, and is reflected over the heart and roots of the vessels. It gives to the heart its smooth shining appearance, and the fluid it secretes lubricates the surface of the organ and allows it to move freely in the pericardium.

The cavities of the heart are lined by a serous membrane similar to that of the blood-vessels.

Blood-Vessels.

There are two sets of blood-vessels, one of which, the arteries, carries the red or oxygenated blood from the heart and distributes it to the different parts of the body; and the other, the veins, carries dark blood, collecting it from the various parts of the body, and returning it to the heart.

The Arteries.

The arteries are composed of three coats, an external, a middle, and a serous one.

The *external* coat is formed of *cellular* tissue, and is firm and strong, not yielding on the application of a ligature.

The *middle* coat is *fibrous* and elastic, having its fibres arranged circularly. It is readily divided by a ligature.

The *internal* coat is a delicate serous membrane, easily torn. Its secretion lubricates the surface of the artery, and facilitates the passage of blood.

The small arteries which ramify on the coats of the arteries to nourish them, are called *vasa vasorum*. The sympathetic furnishes the arteries with nerves.

The Aorta, and its Branches.

The aorta is the main trunk of the arterial system; it arises from the upper and posterior end of the left ventricle, and passing upwards and backwards towards the left side forms a curvature, called its arch, the summit of which is about an inch lower than the upper end of the sternum. Near the origin of the arch there is generally an enlargement or dilatation of the aorta, called its greater *sinus*. After forming its arch the aorta passes to the left side of the spine, about the third or fourth dorsal vertebra, and descends through the thorax to the hiatus aorticus of the diaphragm, through which it passes into the abdomen, and terminates in front of a space between the fourth and fifth lumbar vertebræ, by dividing into two large trunks called the *primitive iliacs*. In its descent through the thorax and abdomen, it is in contact with the left side of the bodies of the vertebræ.

It receives the name of *thoracic aorta,* while passing through the thorax, and *abdominal,* while passing through the abdomen.

The *coronary arteries* are the first branches given off by the aorta; they are distributed to the heart.

The *arteria innominata* is the next branch; it is given off from the arch of the aorta, and after ascending about an inch and a half obliquely towards the right side, divides in front of the trachea into the *right carotid,* and *right subclavian.* The *right carotid* ascends nearly as far as the os hyoides, and divides into the external and internal carotids.

The *left carotid* arises from the arch of the aorta, and passing to the left side of the neck divides, as the right, into the external and internal carotids.

The external carotid is distributed to the more superficial parts of the head and neck, and the internal goes to the brain and eye.

The following principal branches are given off from the *external* carotid, viz. the *superior thyroid* which goes to the thyroid gland; the *lingualis* to the tongue; the facial to the face; the *inferior pharingeal* to the pharynx; the *occipital* to the integuments on the back part of the head; the *posterior auricular* to the integuments of the side of the head; and the *anterior, posterior,* and middle *temporal,* which go to the muscles and integuments on the side and back part of the head.

The *internal maxillary* artery commences at the bifurcation of the external carotid, and winds around the neck of the lower jaw to supply the back portions of the mouth and palate. Its course is very tortuous, and it gives off numerous branches, of which the following are the principal: the *tympanitic,* to the tympanum, or drum of the ear, through the glenoid fissure; the *meningea parva,* to the dura mater through the foramen ovale; the *meningea magna* or *media,* to the dura mater through the spinal foramen; the *inferior dental,* to the teeth through the posterior mental foramen; the *deep temporal,* two in number, which go to the temporal muscle; the *pterygoid* and *buccal* to the muscles and lining membrane of the cheek; the *superior alveolar* or *maxillary* to the *molar teeth, antrum and gums;* the *infra orbital* enters the infra-orbital canal, and is distributed to the canine and incisor teeth, to the antrum, and to the muscles in front of the upper jaw; the *superior palatine* goes to the mouth and palate; the *superior pharyngeal* to the pharynx and Eustachian tube; and the *spheno-palatine* to the lining membrane of the nose, entering through the spheno-palatine foramen.

The vertebral and internal carotid arteries supply the brain; the former enter through the foramen magnum occipitus, and uniting form a large trunk, called the basilar, which passes along the median line of the pons varolii, giving off several branches; the principal being the *superior cerebellar,* and the *posterior cerebral* arteries.

The carotid enters the cranium through the carotid canal, and divides into the *ophthalmic, middle cerebral,* and *anterior cerebral* arteries.

The arteries are united in front and behind by branches called *anterior* and *posterior* communicating, and by this a circle, called the circle of Willis, is formed.

Subclavian Artery.

On the left side the subclavian arises from the arch of the aorta, and on the right from the innominata. The right is consequently shorter, and more superficial than the left.

The subclavian passes out of the thorax over the first rib, and usually gives off five branches, viz.

The *vertebral,* which is the first and largest, and which passes through the foramina of the spinous processes of the six upper cervical vertebræ, and enters the cranium through the occipital foramen.

The *inferior thyroid,* which goes to the thyroid gland, and also gives off the ascending cervical to the muscles of the neck.

The *superior intercostal,* to the two upper intercostal spaces.

The *internal mammary,* which enters the thorax, and passing down over the cartilages of the ribs, gives off branches to the thorax, diaphragm, and abdomen.

The *posterior cervical*, which passes to the back of the neck to be distributed on the muscles there.

After giving off these branches the subclavian artery goes to the axilla or arm-pit, passing between the first rib and the subclavius muscle, where it loses the name of subclavian and takes that of *axillary*.

The *axillary* artery, also usually sends off five branches, though there is much irregularity in this respect.

The *supra-scapular*, to the muscles of the scapula.

The *external mammaries*, mostly four in number, to the shoulder, axilla, and muscles on the front of the thorax.

The *scapular* to the arm-pit, and the muscles on the back of the thorax.

The *anterior circumflex* to the parts in front of the joint; and the *posterior circumflex* to the posterior parts of the joint and the deltoid muscle.

From the axilla to the elbow joint the artery descends at the inner side of the arm on the edge of the biceps flexor muscle, and is called *brachial.*

The *brachial artery* gives off four branches.

The *profunda major* to the outer portions of the arm.

The *profunda minor* to the internal face of the triceps at its lower part, and to the internal condyle.

The *nutritious artery* to the bone through the nutritious foramen.

The *anastomotic*, which passes around the internal condyle, and anastomoses with the ulnar recurrent.

At the elbow joint the brachial artery divides into the radial and ulnar arteries.

The *radial* is the smaller of the two, and except at the wrist is the more superficial. It descends at the outer side of the arm between the tendons of the supinator radii longus and flexor carpii radialis muscles. In its course it gives off the following branches, viz.

The *radial recurrent*, which is distributed about the joint, and anastomoses with the profunda major.

The *superficialis volæ*, which is distributed to the palm of the hand.

The *dorsalis carpi* to the back of the wrist.

The *magna pollicis* to the thumb; this is one of the terminal branches of the radial.

The *radialis indicis* is connected at its origin with the last, and is distributed on the radial side of the fore-finger.

The *palmaris profunda* is the third terminal branch of the radial; it crosses the hand beneath the flexor tendons, forming the arcus profundus, from which branches are sent to the interossei muscles, and uniting at the ulnar side of the hand with the cubitalis manus of the ulnar artery.

The *ulnar artery* is deep seated; it passes from the internal condyle along the inner side of the arm between the tendons of the flexor carpi ulnaris, and flexor sublimis muscles. At the wrist it is superficial, and its pulsations may be distinctly felt; it passes over the annular ligament, and, in the palm, forms the superficial arch.

The following are the branches it sends off:

The *ulnar recurrent*, which is distributed to the muscles of the internal condyle, and anastomoses with the anastomotic.

The *interosseus*, which divides into an anterior and a posterior interosseous branch; the anterior descends the arm, in contact with the interosseous ligament, giving off branches in its course to the deep seated muscles; near the wrist it perforates the ligament, and is distributed to the back of the wrist and hand; the posterior branch soon perforates the interosseous ligament, to be distributed to the extensor muscles of the fore-arm.

The *dorsalis manus* is given off at the lower end of the fore-arm, and distributed upon the back of the hand.

The *superficial arch* (*arcus sublimis*) is formed by the continuation of the radial artery beneath the palmar fascia, from it a branch is sent to the ulnar side of the little finger, followed by three others (digital arteries), each of which on arriving at the heads of their metacarpal bones, divides into two branches, a digito-radial, and a digito-ulnar, to supply the sides of the fingers.

Branches of the Thoracic Aorta.

In descending through the chest the aorta gives off the following branches.

The *bronchial arteries*, generally but one for each lung, though sometimes two or more, pass into the root of the lungs, and are distributed along the ramifications of the bronchia. They are the nutritious arteries of the lungs.

The *œsophageal arteries*, mostly five or six small twigs, arise one after the other from the aorta, and go to the œsophagus.

The *intercostal arteries*, ten on each side, supply the ten inferior intercostal spaces, the two upper spaces being supplied by the subclavian. Those for the right side are longer than the left in consequence of having to cross the spine behind the œsophagus and the vena azygos. Each artery passes along the grooves in the lower margin of the rib for about two-thirds of the length of the latter.

Branches of the Abdominal Aorta.

The aorta in its descent through the abdomen gives off several branches viz.

The *phrenic*, two in number, which go to the diaphragm, and are chiefly distributed on its concave surface.

The *cœliac*, a large trunk, about half an inch in length, which is given off opposite the junction of the last dorsal with the first lumbar vertebra. It divides into three branches—the gastric, hepatic, and splenic.

The *gastric* is the smallest branch and goes to the lesser curvature of the stomach.

The *hepatic* goes to the liver, entering through the transverse fissure; it gives off a branch to the greater curvature of the stomach, called the *right gastro-epiploic;* and another to the gall bladder, called the *cystic*.

Fig. 27.

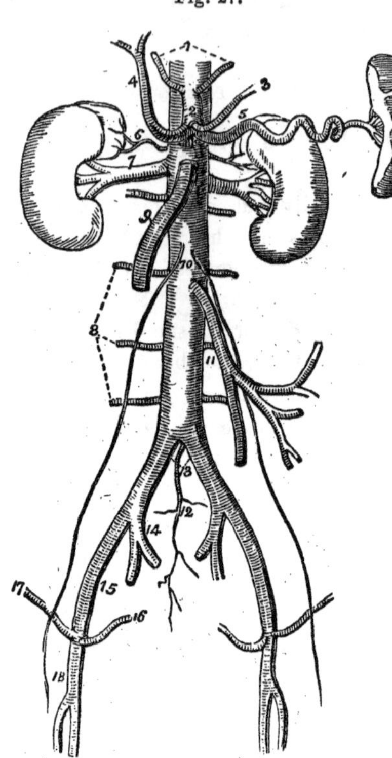

Fig. 27. The abdominal aorta and its branches.

1 Phrenic arteries; 2 cœliac artery; 3 gastric artery; 4 hepatic artery; 5 splenic artery; 6 supra-renal artery of the right side; 7 right renal artery; 8 lumbar arteries; 9 superior mesenteric; 10 spermatic arteries; 11 inferior mesenteric; 12 sacra media; 13 common iliacs; 14 internal iliac of the right side; 15 external iliac; 16 epigastric artery; 17 circumflex iliac; 18 femoral artery.

The *splenic,* which is the largest branch of the cœliac, goes to the spleen; it gives off several small branches to the pancreas, and the *vasa brevia*, and *left gastric* arteries to the left half of the greater curvature of the stomach.

The *superior mesenteric* arises from the aorta about half an inch below the cœliac, and is nearly as large; it is distributed to the whole of the small intestine, and to the right side of the large; three branches go to the latter, viz. the *ileo-colic* to the cœcum, and a portion of the ileum; the *colica dextra* to the ascending colon; and the *colica media* to the transverse colon.

The *capsular arteries,* generally but one on each side, go to the supra-renal capsules.

The *renal* or *emulgent arteries,* mostly one for each side, are large, and go outwards transversely to the kidneys; before entering the fissure of the kidneys they divide into three or four branches.

Fig. 28.

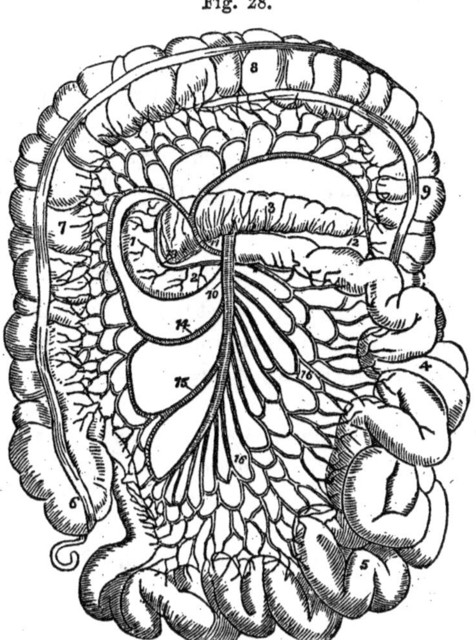

Fig. 28. Distribution of the superior mesenteric artery. 1, 2 Duodenum; 3 pancreas; 4 jejunum; 5 ileum; 6 cœcum and appendix vermiformis; 7 ascending colon; 8 transverse colon; 9 commencement of the descending colon; 10 superior mesenteric artery; 11 colica media; 12 the branch which anastomoses with the colica sinistra; 13 the branch of the superior mesenteric, which anastomoses with the pancreatico-duodenalis; 14 colica dextra; 15 ileo colic; 16, 16 branches to the small intestines.

The *spermatic arteries* arise from the aorta just below the emulgents, or sometimes from the latter, and pursuing a tortuous course, pass through the abdominal rings, and divide into several branches before reaching the testicle. In the female the spermatics are chiefly distributed to the ovaries.

The *inferior mesenteric* usually arises from the aorta about an inch above its termination. It is much smaller than the superior mesenteric, and divides into three principal branches, called the *superior*, *middle*, and *inferior left colic arteries*, which are distributed to the left side of the colon. It also gives off a branch, the *superior hæmorrhoidal*, to the upper part of the rectum.

The *lumbar arteries*, commonly five on each side, correspond with the intercostales, and are distributed chiefly to the loins.

The *middle sacral artery* is a small artery arising from the bifurcation of the aorta, and passing down the sacrum to the coccyx.

The Primitive, or common Iliacs.

The aorta, as mentioned, divides at the fourth lumbar vertebra into the two iliacs, which pass outwards and downwards to the sacro-iliac junction, where they divide into the internal and external iliac.

The *internal iliac*, or *hypogastric*, is a short trunk descending from the sacro-iliac junction into the cavity of the pelvis, and giving off numerous branches, of which the following are the principal:

The *ilio-lumbar*, which is distributed to the loins.

The *lateral sacral*, which divides into four branches to enter the anterior sacral foramina.

The *obturator artery*, which passes out of the pelvis at the obturator foramen, and is distributed by two branches to the obturator and adductor muscles.

The *middle hæmorrhoidal*, which goes to the rectum to the vesiculæ seminales, and to the prostate gland.

The *vesicle arteries* consisting of several branches, which go to the bladder.

The *gluteal artery*, which is a large trunk and one of the terminating branches of the internal iliac, passes out of the pelvis at the upper part of the sacro-sciatic notch, to be distributed by two or three branches to the glutei muscles.

The *ischiatic artery*, which is the anterior of the two terminal branches of the internal iliac, passes out of the pelvis through the sciatic notch, supplying the floor of the pelvis and back of the thigh. Before leaving the pelvis the ischiatic artery gives off a large branch, called the *internal pudic*, which is distributed to the muscles of the perineum and to the penis.

The *internal pudic* gives off several branches, viz.

The *lower hæmorrhoidal*, to the anus and rectum.

The *transversus perineal*, to the muscles and integuments of the perineum. In the lateral operation for stone it is always cut.

The *urethro-vulvar*, to the corpus spongiosum of the penis.

The *dorsi penis*, to the back of the penis.

And the *cavernous artery*, to the corpus cavernosum.

The External Iliac.

The external iliac artery extends from the sacro-iliac junction to Poupart's ligament, under which it passes to the lower extremity, and is then called the femoral artery. Near Poupart's ligament it gives off two branches; one, called the *epigastric*, ascends obliquely upwards and inwards, to be distributed upon the anterior parietes of the abdomen. In its course it passes between the two abdominal rings, and consequently in the operation for hernia is in danger of being cut.

The other, the *circumflex iliac*, also passes obliquely upwards to the crest of the ilium, and is distributed to the muscles of the loins and abdomen.

The Femoral Artery.

This artery is a continuation of the external iliac; it extends from Poupart's ligament down the inner side of the thigh, about two-thirds of its length, where it perforates the tendon of the adductor magnus muscle, and takes the name of *popliteal*. The *femoral* artery sends off the following branches:

The *superficial artery of the abdomen* goes obliquely to the integuments of the lower part of the abdomen, and is there distributed.

The *external pudics*, two or three small arteries which go to the integuments.

The *profunda*, a large artery going to the muscles of the upper part of the thigh. It gives off a large branch called the *external circumflex,* which supplies the muscles on the outside of the thigh; and another, called the internal circumflex, to the muscles on the inside of the thigh. Several smaller branches, called perforating arteries, are given off by the profunda and the internal circumflex.

The *anastomic artery* is sent off by the femoral, and descending to the knee anastomoses with the internal articular arteries.

The Popliteal Artery.

The popliteal is a continuation of the femoral artery, and extends from the adductor tendon to the opening in the interosseous ligament below the head of the tibia, where it divides into the anterior and posterior tibial.

It gives off the *superior, middle,* and *inferior* articular arteries to the knee joint; and the *gemellar,* two in number, to the heads of the gastrocnemius muscle.

The Anterior Tibial.

The anterior tibial artery perforates the interosseous ligament, and descends the leg in front of it to the ankle joint, giving off in its course numerous branches, as follows:

The *recurrent tibial,* which goes upwards, and anastomoses with the arteries about the knee joint.

The *internal malleolar,* to the inner side of the ankle joint.

The *external malleolar,* to the outer side of the ankle joint.

The *tarsal,* to the external ankle and to the tarsus.

The *metatarsal,* to the toes by three branches. It forms an arch at the roots of the metatarsal bones.

The *dorsal,* to the outer side of the great toe, and the inner side of the second toe.

And the *pedal,* a branch passing through the first interosseal space to the sole of the foot to join the external plantar artery.

The Posterior Tibial.

This artery descends from the head of the tibia to the os calcis in a line from the middle of the ham to the external ankle. It is beneath the muscles of the calf of the leg.

The *peroneal* arises from the posterior tibial, and dividing into several branches is distributed to the muscles on the back of the leg.

On reaching the os calcis, the posterior tibial artery divides into the internal and external plantar arteries.

The *internal plantar* is a small artery, which passes along the inner margin of the foot, and is distributed to the great toe.

The *external plantar* is considerably larger than the last, and passes obliquely across the sole of the foot to the outer margin, forming an arch from which branches (*digital arteries*) are given off to the interosseous spaces, and to the toes.

The Veins.

The veins collect the blood from the different parts of the body, and convey it to the heart. They are much more numerous than the arteries; the deep seated arteries generally having two accompanying veins called *venæ comitas.* All the venous blood is emptied into the right auricle through two large trunks, called the ascending and descending vena cava. The veins of the head, upper extremities, and thorax, unite to form the descending vena cava.

The parietes of the veins, like the arteries, are formed of three coats; these coats, however, are much thinner than those of the arteries, which render the veins flaccid when empty and easily distinguishable from the arteries, the elastic middle coat of which enables them to retain the cylindrical shape.

The veins are likewise furnished with numerous valves, formed of two or three crescentic folds of the lining membrane, which open towards the heart.

In the brain and the bones the canals or channels which convey the venous blood, are termed *sinuses*, and are lined by the internal coat of the veins.

Veins of the Head and Neck.

Most of the veins of the head and neck have the same names as the arteries they accompany.

The internal maxillary, occipital, temporal, and other of the principal trunks join together to form the *external jugular*, which is a superficial vein extending from the parotid gland to the subclavian vein, into which it empties behind the outer end of the clavicle. The external jugular is covered externally by the skin, superficial fascia, and platysma-myoides muscle.

The *internal jugular vein* is larger than the external; it extends from the base of the occiput to the posterior face of the clavicle near its sternal end, where it unites with the subclavian to form the vena innominata. It receives the blood from the sinuses of the cranium, and from some of the superficial veins.

Fig. 29.

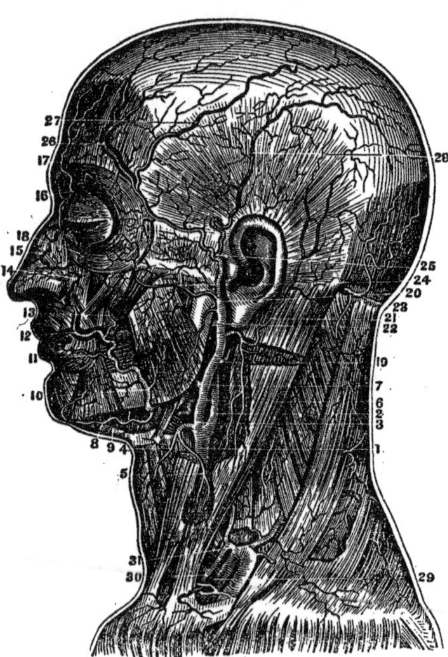

Fig. 29. A lateral view of the veins of the head and neck. 1 Vena innominata; 2 internal jugular; 3 external jugular; 4 superior thyroid; 5 inferior thyroid; 8 submental; 10 inferior maxillary; 11, 12 superior and inferior labial veins; 13 angularis nasi; 14 anterior facial; 15 nasalis externus; 16 supra orbital; 18 infra orbital; 19 fascialis posterior; 21 occipital vein; 23 auricularis posterior; 25 temporalis superficialis; 26, 27, 28 temporalis anterior, middle, and posterior.

Sinuses of the Dura Mater.

The channels, or passages, by which the blood is removed from the brain, are formed between the lamina of the dura mater, and termed *sinuses.*

The principal are:

The *superior longitudinal sinus,* a channel of a triangular shape, commencing at the foramen cœcum by a small vein from the root of the nose, and gradually increasing in size forms an arch in the middle line of the arch of the cranium, which terminates at the internal occipital cross in an enlargement called the *torcular herophili,* where it receives the blood from other sinuses.

The *inferior longitudinal sinus,* is situated between the lobes of the cerebrum in the falx cerebri; it also forms an arch, and empties into the straight sinus at the anterior edge of the tentorium.

The *straight* or *fourth sinus,* extends along the median line of the tentorium, from the falx cerebri to the torcular herophili.

The *vena azygos,* is a short trunk formed by the *venæ Galeni,* which empties into the fourth sinus.

The *lateral sinuses,* on either side, commence at the torcular herophili, and pass over the occipital, parietal, and temporal bones to the posterior foramen lacerum, through which they pass out of the cranium, and form the commencement of the internal jugular vein.

The *circular sinus,* surrounds the pituatary gland in the sella turcica, communicating on either side with the cavernous sinuses.

The *cavernous sinuses,* receive the blood from the ophthalmic veins; they are venous cells of a spongy structure, situated on either side of the sella turcica.

The *superior* and *inferior* petrosal sinuses, are small channels, traversing the petrous portion of the temporal bone to empty into the lateral sinus.

The *anterior, occipital sinus,* passes across the basilar process of the occipital bone, forming a communication between the two inferior petrosal sinuses.

The *posterior occipital sinus* extends from the torcular herophili in the lower edge of the falx cerebri to the foramen magnum, where it divides and empties into the lateral sinus on either side.

The name of *emissaries of Santorina* has been given to the small veins, which pass through the minute foramina in the bones of the cranium, to communicate with the sinuses of the brain.

The Veins of the Upper Extremities.

These veins are superficial and deep seated; the latter accompany the arteries, and take the same name; two veins usually accompany one artery.

The superficial veins lie between the skin and the brachial aponeuroses. They anastomose frequently with each other, commence generally on the back of the fingers, and finally unite into two principal branches, called the cephalic and basilic vein.

The *cephalic vein* commences on the thumb, fore-finger, and back of the hand, and passes up the radial side of the fore-arm to the elbow, where it is joined by the median cephalic, and pursuing its course up the outer side of the arm, empties into the subclavian vein beneath the clavicle.

The *basilic vein* commences on the little finger, and passes up the ulnar side of the fore-arm to the elbow, where it receives the median basilic, and continuing its course up the arm along the inner margin of the biceps muscles, where it is joined by the venæ comites, and becomes the axillary vein.

The *median vein* collects the blood from the palm of the hand, wrist, and front of the fore-arm, and ascends in front of the fore-arm to within a few inches of the elbow joint, where it divides into two branches, one of which called the *median cephalic* runs outwardly to join the cephalic vein; the other, the *median basilic*, goes inwards to join the basilic vein.

The *axillary vein* is formed by the union of the basilic and the brachial veins; it passes up the arm in the same sheath with the axillary artery to the axilla, where, like the artery, it takes the name of subclavian.

The *superior* or *descending vena cava* is formed by the junction of the right and left *innominata;* it empties into the right auricle of the heart.

The Veins of the Lower Extremities.

The veins of the lower extremity, like those of the upper, are superficial and deep seated, and the latter generally take the name of the artery they accompany.

The *popliteal vein* commences behind the head of the tibia, and extends upwards along with the popliteal artery to the perforation in the tendon of the adductor magnus muscles, after which it takes the name of femoral.

The *femoral vein* is a continuation of the popliteal, and passes upwards along with the artery of the same name to Poupart's ligament, where it becomes the external iliac vein.

The *small* or *external saphena vein* is superficial, commencing by the union of several small branches at the outer side of the top of the foot and ankle, and passing up the posterior and outer side of the leg to the ham, empties into the popliteal vein.

The *great* or *internal saphena vein* is also superficial; it commences by a number of roots from the sole, and the inner and upper part of the foot, and ascending along the internal face of the leg, and thigh to within about an inch and a half of Poupart's ligament empties into the femoral vein.

Veins of the Abdomen.

The principal of these are the external and internal iliacs; the spermatics of the testicles; the renales or emulgents from the kidneys; the hepatic in three branches, from the liver, all of which empty directly into the ascending vena cava; the vesical plexus from the bladder, and the uterine plexus from the uterus, which empty into the internal iliac; and the hæmorrhoidal veins from the rectum, which empty into the inferior mesenteric.

The *inferior* or *ascending vena cava* commences between the fourth and fifth lumbar vertebræ, by the union of the iliac veins, and ascending along the spinal column at the right of the aorta, receives in its course the abdominal veins, and passing through the diaphragm into the chest, empties into the right auricle of the heart.

The Portal Vein.

The portal vein collects the blood from all the viscera of the abdomen, and entering the transverse fissure of the liver, is included with the biliary duct and hepatic artery, in a common investment of cellular tissue, called the capsule of Glisson; it ramifies through the substance of that organ previous to entering the general circulation.

The veins which contribute to form the portal are: the superior and inferior mesenteric, the splenic, the gastric, and the pancreatic; the trunk of the portal vein, which is about four inches in length, extends from the posterior face of the pancreas to the transverse fissure of the liver.

Peculiarities of the Fœtal Circulation.

The circulation of the fœtus, in consequence of the absence of respiration, differs very materially from that function in the adult. For its nutrition and development an alliance through the circulatory organs of the mother is necessary.

The peculiarities of the fœtal circulation consist in: the *ductus venosus,* a vein leading from the umbilical vein along the margin of the liver to the ascending vena cava; the communication of the right and left auricles through the *foramen ovale;* and the *ductus arteriosus,* a branch from the pulmonary artery, which conducts the blood returned from the head into the aorta just behind the origin of the left subclavian artery. When the current of the blood is changed after birth, by respiration, the ductus venosus and arteriosus, shrivel up into ligamentous cords, and the aperture between the auricles is closed by the adhesion of its valve.

The *placenta* is the organ through which the effete blood of the fœtus is regenerated, or ærated. It is closely attached to the uterus of the mother, and the change in the blood is probably effected by interstitial circulation, as there appears to be no direct connection of blood-vessels between the mother and fœtus. The umbilical cord extends from the placenta to the umbilicus of the child, and consists of one *umbilical vein* and two *umbilical arteries;* the latter are twisted around the former, and conduct the blood to the placenta; the former conveys the ærated blood back again to the child. Thus it will be observed, that the umbilical vein and the ductus venosus carry *arterial* blood; and the umbilical arteries and the ductus arteriosus carry *venous* blood.

Commencing at the placenta, the circulation of the fœtus pursues the following course. The blood passes along the umbilical vein to the umbilicus of the child; it then penetrates the abdomen, after which it divides into two currents, one of which goes through the ductus venosus into the ascending vena cava; the other passes through the vena portarum to the liver. The portion sent to the liver, after ramifying through that organ, is collected by the hepatic veins, and likewise discharged into the ascending cava.

The contents of the ascending cava are emptied into the *right* auricle of the heart, from which it passes into the *left* auricle through the ostium venosum. From the left auricle it passes into the left ventricle, by which it is distributed by means of the aorta throughout the system. That portion of the blood which goes to the head and upper extremities, is collected in the descending vena cava, and also discharged into the *right* auricle, from whence it passes into the right ventricle, from which a small portion goes into the pulmonary artery, and the remainder into the aorta through the ductus arteriosus.

It will be perceived, therefore, that all parts of the body receive mixed arterial and venous blood, with the single exception of the liver, to which organ the blood is sent from the placenta unmixed.

The umbilical arteries are continuations of the internal iliacs; they pass up at the sides of the bladder and go to the navel, from which point, as mentioned, they form part of the umbilical cord. After the change in the circulation, at birth, they become the round ligaments of the bladder.

The Lymphatics.

The lymphatics, so called from the transparent color of the fluid they carry, are found in all parts of the body; their office is that of interstitial absorption, by which the effete parts of the body are removed, and room made for new depositions. They are small, transparent, cylindrical vessels, rarely exceeding a line in diameter. Like the arteries, they have three coats, and when distended they present a knotted appearance, which is owing to their being furnished with valves and sinuses.

In the intestines the lymphatics, which commence in the villous coat, are frequently called *lacteals*, and contain the chyle.

The lymphatic is also spoken of as the absorbent system.

Numerous lymphatic glands exist in the mesentery, groin, axilla, and neck, into which the lymphatic vessels enter, and pass out again. They form frequent anastamoses with each other, and become larger and less numerous, as they proceed from their origins.

All the lymphatic vessels from the various parts of the body collect into two large trunks, one of which is placed on each side of the body.

The trunk on the right side receives the lymphatics from the right side of the head and neck, and the right upper extremity and lung; it is short, and discharges its contents into the venous system at the junction of the right internal jugular and subclavian vein. The trunk on the left side, called the *thoracic duct*, receives the contents of the lymphatics and chyliferous vessels of the rest of the body; it is the principal lymphatic vessel of the body, and commencing in the abdomen with an enlarged extremity, called *receptaculum chyli*, which receives the contents of the lacteals and the lymphatic vessels of the lower extremities, it passes into the thorax through the diaphragm, and ascending through the posterior mediastinum in front of the spinal column, along with the aorta, empties its contents at the junction of the left subclavian and internal jugular veins.

Physiology of the Circulation.

By the circulation of the blood is understood that function by which the arterial blood, fitted for nutritive purposes in the lungs, is distributed to every part of the body. The organs and vessels, by which this function is accomplished, viz. the heart, arteries, veins, and the capillary vessels, intermediate between the arteries and veins, constitute the vascular system.

The heart is the centre of the circulation, and may be considered as a double organ, or as two distinct hearts brought together for convenience of package,—the right one containing black or venous blood only, and the left one red or arterial blood only. The circulation from the two hearts, or the two sides of the heart, are entirely distinct; that from the right is called the lesser or pulmonic circulation, and that from the left, the greater or systemic.

Commencing at the heart, the route of the circulation is as follows; the venous blood from all parts of the body is collected in the *right auricle*, from which it passes through the auriculo-ventricular orifice into the *right ventricle;* thence it is sent through the pulmonary artery to the lungs to be ærated, after which it is collected by the pulmonary veins, and carried to the *left auricle*, out of which it passes through the auriculo-ventricular orifice into the *left ventricle*, from whence it is distributed by means of the aorta throughout the system.

The chief propelling power of the circulation is the alternate contraction and dilatation of the heart. This motion of the heart is constant and unremitting during life; its cessation, even for a short time, terminating existence. It is independent of the nervous system, and also of the stimulus of the blood, as it continues after all nervous communication is cut off, and when the heart is empty, and in fact after it has been removed from the body. The ventricles are the principal agents in the propulsion of the blood, for which purpose they have strong muscular parietes; the auricles have but slight contractile powers, their chief use being as receptacles. The contraction of the auricles and ventricles is not synchronous. The two auricles act together, as do also the two ventricles. The contraction of the auricles which forces the blood into the ventricles, is immediately followed by the contraction of the ventricles, during which the auriculo-ventricular valves are closed to prevent the regurgitation of the blood into the auricles, and consequently it passes into the aorta and pulmonary artery. Dilatation of all the cavities succeeds their contraction. The term *systole* has been given to the contraction of the heart, and that of *diastole* to its dilatation. The pulse is caused by the projection of the blood into the arteries from the ventricle, and corresponds with the contraction or systole. The contractile nature of the muscular coat of the arteries aids in the propulsion of the blood, and also in the production of the pulse. In the capillary and venous circulation there is no pulse.

By the time the blood reaches the capillaries, the force of the heart is lost. Of the circulation carried on in this system of vessels but little more can be said than that it is due to vital action. The nutritive properties of the arterial blood are here yielded up, and the effete portions enter the veins to be returned to the heart and lungs, there to be given off, or again vitalized.

The forces which aid in returning the blood to the heart are: the suction power of the heart, muscular motion, and inspiration. When the heart dilates, and when the chest is expanded by the descent of the diaphragm, a vacuum is created, which the blood rushes towards the chest to fill.

At every contraction of a muscle, the veins of the part are pressed upon, and their valves allowing the flow of the blood but in one direction, towards the heart, it must consequently be driven on in that direction.

The muscular force of the heart has been variously estimated by different observers, some allowing it to be equal to but ten or fifteen pounds, other estimating it at thirty or forty. A true estimate is of course exceedingly difficult to arrive at, as it is greatly modified by age, sex, temperament, and a variety of causes.

The usual number of contractions of the heart, or pulsations, per minute, in the adult is from seventy to seventy-five. In youth they are more frequent, and in old age less so.

In the female sex the pulse is also more rapid than in the male. Temperament, muscular exertion, mental emotion, &c., all exercise a controlling influence over the action of the heart.

On applying the ear over the region of the heart, two distinct sounds are perceptible, during each beat of the heart. The first is a dull, lengthened sound; the second quick and sharp; they follow each other in quick succession, and are succeeded by a short interval of repose, after which they recur again, and so on. The first is synchronous with the contraction, and the second with the dilatation of the ventricles. About one-half the whole period between the commencement of one pulsation, and the commencement of the next is occupied by the first sound; one-fourth by the second; and the remaining fourth is the period of repose. The causes of the first sound are: the rush of the blood through the orifices of the aorta and pulmonary artery; the passage of the blood over the rough internal surfaces of the ventricles; the flapping of the auriculo-ventricular valves; the sound of muscular contractions; and the impulse of the heart against the chest.

The second sound is caused by the closure of the valves, at the mouths of the aorta and pulmonary arteries.

At each contraction of the heart, it is projected forwards, and strikes against the parietes of the thorax, in the region of the fifth and sixth ribs; the shock this occasions is called the *impulse* of the heart.

There is some variation in the capacity of the cavities of the heart, the right auricle and ventricle being somewhat more capacious than the left.

Tht circulation was discovered by *William Harvey* of London, in 1719.

Nervous System.

In this system are included the brain, spinal marrow, and nerves.

The material or *tissue* of which it is composed, is called *neurine*, and is of a soft, pulpy consistence.

This tissue consists of two portions, one of which is white or *medullary*, and of fibrous structure; the other gray or *cineritious*, and globular in structure.

The nerves are composed of the medullary substance, and consist of parallel fasciculi or bundles of fibres, capable of being subdivided into filaments. Each nerve, as well as each particular fibre, is enveloped in a sheath, called the *neurilemma*.

A *ganglion* is a knot occurring in the course of a nerve, and by which they obtain an increase of volume and power. Ganglia are of different sizes and shapes, and consist of a union of white and gray matter.

An *anastomose* is the junction of the filaments of the same nerve, or of different nerves.

A *plexus* is the junction or interchange of the larger fasciculi of the same nerve, or of different nerves, forming a network.

The Spinal Marrow.

The spinal marrow is contained within the vertebral canal, and extends from the atlas or first vertebra of the neck to the first or second lumbar vertebra. Its general form is cylindrical, though it is somewhat flattened in front and behind, and has an enlargement in the neck and loins. Its diameter, with the addition of its membrane, is much smaller than that of the spinal canal, by which provision injury from pressure is guarded against.

It is divided, longitudinally, into two symmetrical parts, by an anterior and a posterior fissure. The posterior fissure is somewhat deeper than the anterior. It has also a lateral fissure on each side, placed somewhat posterior to the middle, and passing inwards and forwards.

The spinal marrow is composed of medullary and cineritious

matter, the former being internal, and the latter external; the cineritious matter is also much more abundant than the medullary.

From the sides of the spinal marrow, thirty pairs of nerves are sent out, which, like the vertebræ, are divided into cervical, dorsal, lumbar, and sacral. Of these, eight pairs belong to the neck, twelve to the thorax, five to the loins, and five to the sacrum. Sometimes there is an additional pair, making thirty-one in all.

Each nerve arises by two roots, one of which comes from the anterior cord of the spinal marrow, and the other from the posterior cord. The posterior root is larger than the anterior, and upon it is a ganglion; the roots are separated by a process of pia mater, called the *ligamentum denticulatum*, and after penetrating the dura mater by separate foramina, unite to form a single trunk.

The lumbar and sacral pairs arise from the lower extremity of the spinal marrow, and form a cluster resembling the tail of a horse, hence the name of *cauda equina* has been applied to them.

Three membranes envelop the spinal cord; one, the external, is called the *dura mater;* another, the internal, is the *pia mater*, and the third, which is between the others, is the *arachnoidea.* The *dura mater* is continuous with that of the brain, and terminates below in a closed extremity; it is a white fibrous membrane, affording a loose investment to the spinal canal, except at the first cervical vertebra, to which it adheres firmly. Between it and the parietes of the canal is placed a quantity of loose cellular tissue containing fat and serum.

The *tunica arachnoidea* is placed next to the dura mater; it is very thin and transparent.

The *pia mater* adheres closely to the spinal marrow; it is a cellular membrane, made up almost entirely of blood-vessels; it sends processes into the anterior and posterior fissures of the medulla spinalis; below it terminates in round cord-like processes continuous with the roots of the nerves constituting the cauda equina. It is also continued along the nerves constituting their neurilemma or sheath.

The Brain.

The brain is contained within the bones of the cranium, and is oval in shape.

It is composed of cineritious and medullary matter, and consists of four principal parts, viz. the *medulla oblongata,* which is a continuation of the spinal marrow, or its superior part; the *pons Varolii,* called also *protuberantia annularis,* which is placed at the superior extremity of the medulla oblongata; the *cerebrum,* which occupies seven-eighths of the cavity of the cranium; and the *cerebellum,* which is situated at the base of the cranium.

Like the spinal marrow, the brain also is enclosed in three membranes,—the dura mater, tunica arachnoidea, and pia mater.

The dura mater is the most external, lining the whole of the cavity of the cranium, to the bones of which it is firmly attached. Its structure is fibrous. It consists of two lamina, from the internal of which several processes are formed. One of these, the *falx cerebri,* is situated under the middle line of the head, and extends from the crista galli in front to the tentorium behind, separating the hemispheres of the brain. Its breadth in front is about an inch, and behind about two, or two and a half inches. Another process is called the *tentorium,* and like the last is crescentic in shape; it is stretched horizontally across the cranium, separating the cerebellum from the posterior lobes of the brain. In front it is continuous with the falx cerebri. The *falx cerebelli* is a small triangular process, extending from the lower surface of the tentorium to the foramen magnum; it separates the two lobes of the cerebellum.

The *tunica arachnoidea* is placed between the dura mater and pia mater; it is a thin, transparent, serous membrane, and adheres closely to the pia mater.

The *pia mater* is a cellular membrane, placed next to the substance of the brain, and extending to the bottom of the fissures between the convolutions, consequently its internal surface is very irregular; its external surface, being in contact with the arachnoid membrane is smooth and shining.

This membrane is made up almost entirely of blood-vessels. Clusters of small, white, granular bodies, called *glands of Pacchioni* are contained in its meshes, the use of which is not known.

The Medulla Oblongata.

The medulla oblongata, as mentioned, is the upper part of the spinal cord; it extends from the margin of the atlas to the pons Varolii, and is about an inch in length, and three-fourths of an inch in breadth at the base; it gradually increases in size as it ascends. Like the medulla spinalis, it is divided by an anterior and a posterior fissure into symmetrical halves, and each half consists of three portions, viz. the *corpora pyramidalia,* two cord-like cylindrical portions, one on each side of the anterior fissure, and united at their lower extremities by an interchange of fibres. The *corpora olivares,* are two oval bodies situated posterior to the pyramidalia and separated from them by a fissure. The *corpora restiformia* are elliptical elevations, situated at the posterior part of the medulla oblongata, and separated from each other by the posterior fissure. They are about one inch in length, and united below by transverse fibres of medullary matter.

The Pons Varolii.

The pons Varolii, or annular protuberance, is a large cuboidal mass of medullary matter, placed at the top and in front of the medulla oblongata, and resting on the basilar process at the base of the cranium; its internal fibres are longitudinal, being a continuation of those of the medulla oblongata; externally they are transverse. A superficial fossa divides it into two symmetrical halves. Four *crura* proceed from it, two to the cerebrum, and two to the cerebellum.

The Cerebellum.

The cerebellum is situated in the posterior fossa of the cranium, beneath the posterior lobes of the cerebrum, and separated from it by the tentorium.

It is oblong and flattened, having its long diameter transverse, and constitutes about one-sixth part of the brain. It is divided by a longitudinal fissure into two lobes or hemispheres. On making a vertical section through one of the lobes, an arborescent arrangement is presented, caused by the intermingling of the gray and white matter, to which the name of arbor vitæ has been given. A fasciculus of medullary matter, called the *crus cerebelli*, connects each lobe of the cerebellum with the pons Varolii; another fasciculus, called the *valve of the brain*, extends from the corpus restiforme of the medulla oblongata to the under surface of the cerebellum.

Externally the cerebellum is formed of cineritious, and internally of medullary matter.

The Cerebrum.

The cerebrum is a large ovoid mass, six or seven times as large as the cerebellum, and weighing from three to four pounds. It is divided into two parts or hemispheres by the deep fissure above (superior longitudinal fissure), and each hemisphere is subdivided on its under surface into three lobes—anterior, posterior, and middle. The anterior and middle lobes rest upon the anterior and middle fossa of the cranium, and the posterior lobes upon the tentorium. The fissure of Sylvius is between the anterior and middle lobes. The surface of the cerebrum consists of a number of convolutions or gyri, separated from each other by deep fissures (sulci), which give it an exceedingly irregular, tortuous appearance.

The periphery of the convolutions to the depth of about a fourth of an inch is composed of cineritious substance, and the interior of medullary substance.

The longitudinal fissure completely separates the two anterior lobes from each other, extending between them to the base of the brain; the posterior lobes are also separated in the same manner. But between the middle lobes, a broad arched band of medullary matter, called the *corpus callosum*, extends from side to side at the bottom of the longitudinal fissure, connecting the two hemispheres. This band is mostly composed of transverse fibres. The *crura cerebri*, coming from the anterior margin of the pons Varolii, also pass from the middle fissure to each hemisphere. They are two thick, cylindrical, white cords of longitudinal fibres, which terminate in the convolutions after expanding in all directions so as to constitute the principal part of the hemispheres. Between the crura are two white globular bodies, about as large as a pea, called the *eminentia mammillares*. In front of these is a soft mass of cineritious matter called the *tuber cinereum*. The term *infundibulum* has been given to a reddish, hollow, conical body, the base of which is on the tuber cinereum and the apex, extending to the *pituitary gland*. This latter is a light colored, vascular body, consisting of two lobes situated in the sella Turcica. A triangular arch of medullary matter, the base of which is continuous with the posterior part of the corpus callosum, and the apex joined to the eminentia mammillares by two crura, is called the *fornix*. A vertical septum, consisting of two lamina, and separating the lateral ventricles, having its upper extremity attached to the fornix, and its lower to the corpus callosum, is called the *septum lucidum*.

The *velum interpositum* is a reflection of the pia mater immediately beneath the fornix; the *plexus choroides* is a net-work of veins contained in its edges. The *pineal* gland is a small, reddish, conical body, placed upon the tubercula quadrigemina, and connected with the optic thalamus; it often contains particles of calcareous matter. The ancients imagined it to be the seat of the soul. The *tubercula quadrigemina* are four prominences situated on the upper part of the crura cerebri, and behind the optic thalamus; a passage under them is the aqueduct of Sylvius. The *thalami optici*, two in number, are situated on the superior face of the crura cerebri; they are convex above and internally, and are composed of a mixture of medullary and cineritious matter. Their posterior extremities have three rounded prominences, called *corpora geniculata*.

The *corpora striata* are two oblong masses of gray matter,

situated in front of the thalami optici, and at the bottom of the lateral ventricles.

In the brain are *five ventricles*, or cavities; two of them are called lateral, and the others, the third, fourth and fifth.

The *lateral ventricles* are situated in the centre of the hemispheres, and are separated from each other by the septum lucidum. They are horizontal and very irregular in shape; the roof is formed by the corpus callosum, and the floor by the fornix, thalami optici, and corpora striata. Each contains three depressions, called *cornua*, of which one is anterior, one posterior, and one inferior; an oblong eminence on the inner side of the posterior cornua is called *hippocampus minor*, or ergot; and a ridge on the floor of the inferior cornua is called *hippocampus major*. The lateral ventricles communicate with each other, and with the third ventricle by the foramen of Monro.

The *third ventricle* is situated between the thalami optici; it is a narrow, oblong cavity bounded above by the velum interpositum and the fornix, and below by the tuber cinereum, crura cerebri, and eminentia mammillares. It communicates with the lateral ventricles by the foramen of Monro, and with the fourth by the aqueduct of Sylvius.

The *fourth ventricle* is an irregular triangular cavity, situated between the pons Varolii, cerebellum, and medulla oblongata; above, it is bounded by the valve of the brain and the tubercula quadrigemina; its floor is formed by the *calamus scriptorius*. It communicates with the third ventricle.

The *fifth ventricle* is placed between the lamina of the septum lucidum, and has no communication with the other ventricles.

Nerves of the Cranium.

There are nine pairs of nerves arising from the cranium, which are designated numerically, as well as by their function or distribution.

The *olfactory* nerve, or *first* pair arises by three roots from the base of the brain at the corpora striatum which unite in the fissure of Sylvius. It passes forward, converging towards its fellow, to the cribriform plate of the ethmoid bone, where it forms a large, soft bulb from which it sends filaments into the nose to supply the Schneiderian membrane.

The *optic nerve* or *second pair*, arises by a single broad root from the thalamus opticus and the tubercula quadrigemina, and going forward to the anterior part of the third ventricle, forms a junction with its fellow in the form of the letter X, called the chiasm or crossing of the optic nerves, after which it enters the orbit of the eye to join the retina.

The *motor occuli*, or *third pair*, arises from the crus cerebri, and passing into the orbit through the sphenoidal foramen, is distributed to most of the muscles of the eye-ball.

The *patheticus*, or *fourth pair*, arises by two roots from the valve of the brain, and entering the orbit through the sphenoidal foramen, is distributed to the superior oblique muscle of the eye-ball. It is the smallest of the nerves, coming from the encephalon, and is not larger than a sewing thread.

The *trifacial, trigeminus*, or *fifth pair*, arises by three roots, from the medulla oblongata, and emerging from the side of the pons Varolii, enters a canal of the dura mater at the fore part of the petrous portion of the temporal bone, where it forms a ganglion, called the ganglion of *Gasser*, from which proceed three branches, viz. the ophthalmic, superior maxillary, and inferior maxillary.

The ophthalmic branch emerges through the sphenoidal foramen, and is distributed to the orbit, lachrymal gland, and integuments and muscles of the forehead.

The superior maxillary branch passes through the foramen rotundum, and is distributed to the upper jaw and face.

The inferior maxillary branch emerges at the foramen ovale, and is distributed to the tongue, and to the muscles and teeth of the lower jaw.

The *motor externus oculi*, or *sixth pair*, arises from the corpus pyramidale by two roots, and passing forward through the cavernous sinus, it enters the orbit through the sphenoidal foramen, and is distributed to the abductor occuli muscle.

The *seventh pair* includes the *facial* and *auditory* nerves, and arises from the corpus restiforme and calamus scriptorius. Both branches enter the internal meatus. The fascial, which is likewise called *portia dura,* passes out through the stylo-mastoid foramen, and penetrating the parotid gland, is distributed by numerous branches to the face. The auditory is also called the *portio mollis.* It is distributed to the internal ear.

The *eighth pair* includes the glosso-pharyngeal, the pneumo-gastric, and the spinal accessory. It arises by filaments from the corpus olivare, medulla oblongata, and the medulla spinalis. The *glosso-pharyngeal* passes through the posterior foramen lacerum, and is distributed to the side and root of the tongue, and to the tonsils and pharynx. The *pneumo-gastric* also passes out through the posterior foramen lacerum, and descending the neck, included in the sheath with the vessels, enters the thorax and is distributed to the lungs and stomach. The spinal accessory also emerges from the posterior foramen lacerum, and is distributed to the muscles and integuments of the neck.

The *hypo-glossal* nerve, or *ninth pair,* arises by several fasciculi from the medulla oblongata, and passing through the posterior condyloid foramen of the os occipitus, is distributed to the muscles of the tongue.

The Spinal Nerves.

The nerves arising from the spinal cord are divided into the cervical, dorsal, lumbar, and sacral. As mentioned in the account of the spinal cord, they arise by two roots, one coming from the anterior portion of the cord, the other from the posterior portion—the anterior is the motor root, the posterior the sensitive root. After the union of the roots in the intervertebral foramen, the spinal nerves divide into two trunks, the *posterior* of which are much the smaller and go to the muscles of the back; the *anterior* are large, and uniting with the ganglions of the sympathetic, form plexuses, from which the principal nerves of the muscles of the trunk and extremities are derived.

Cervical Nerves.

The first nerve given off from the medulla spinalis, is called the *sub-occipital;* it passes out between the occiput and atlas; its anterior fasciculus is the smaller, and following the course of the vertebral arteries, is partly distributed to the muscles on the front of the vertebræ, and partly joins the pneumogastric and hypoglossal nerves, and the cervical ganglion of the sympathetic; its posterior fasciculus is distributed to the muscles on the back of the neck. The next seven are the cervical nerves proper, of which the three superior anastomose freely with each other, and form a *cervical plexus* at the side of the neck, from which numerous branches are sent to the muscles and skin of the neck. The *phrenic* nerve, also, arises from this plexus, and passing down through the anterior mediastinum, divides into several branches, and is distributed to the diaphragm.

The four inferior cervical nerves and the first dorsal, after sending off filaments to the sympathetic, unite to form in the axilla the *brachial plexus.* From this plexus the following branches are given off which supply the shoulder, axilla, and upper extremity. The *scapular* branch goes backwards, passing through the coracoid notch, and supplies the muscles of the shoulder and scapula. It is quite small.

The *subcapular* and *thoracic* branches, generally five or six in number, supply the muscles of the parietes of the thorax and those under the shoulder.

The *circumflex,* or *axillary* branch, winds around the humerus in company with the posterior axillary artery, and is distributed to the deltoid muscle. The *internal cutaneous* and the *external musculo-cutaneous* are two small branches, which pass down the arm, and are distributed to the muscles and integuments of the fore-arm.

The *radial nerve,* also called musculo-spiral, winds around the humerus, passing between the heads of the triceps muscle, which it supplies with branches, and then descends to the fore-arm and is spent upon its muscles, and upon the wrist and thumb. The *radial* nerve also descends to the fore-arm, passing under the internal coudyle, and is distributed to the teguments on the ulnar side of the back of the hand, the little finger and the ulnar side of the ring finger. The *median* nerve is the largest of the branches given off by the brachial plexus; it descends the arm at the inner edge of the biceps muscle, along with the brachial artery; on reaching the elbow it passes between the heads of the pronator teres muscle, and descending the fore-arm between the flexor sublimus and profundus muscles, giving off branches in its course, is finally distributed to the thumb, and to the fore, middle, and one side of the ring finger.

Dorsal or Thoracic Nerves.

There are twelve pairs of these; the first pair passing out through the intervertebral foramen between the first and the second dorsal vertebræ, and the twelfth between the last dorsal and first lumbar vertebra. The posterior branches go backwards between the transverse processes of the vertebræ, and are distributed to the muscles on the back of the thorax and those lying along the spine. The anterior branches which are the larger, pass along the intercostal grooves, and are distributed to the muscles of the thorax.

The first dorsal nerve sends a branch to the axillary plexus. The second and third send branches, called *intercostal humeral*, to be distributed upon the integuments of the arm.

Lumbar Nerves.

Of these there are five on each side; the posterior branches are small, and go to supply the muscles of the loins; the anterior branches unite to form the *lumbar plexus*, which is placed between the psoas magnus and quadratus lumborum muscles. From this plexus a number of branches are given off: two or three of these are quite small, and go to the muscles of the abdomen. The *external spermatic* branch goes to the groin, and supplies the glands and the cremaster muscle. The *external cutaneous* goes to the commencement of Poupart's ligament, at the commencement of which it emerges from the abdomen, and is distributed to the integuments and muscles on the inside of the thigh.

The *anterior crural* is the largest branch arising from the lumbar plexus; it passes out of the abdomen under Poupart's ligament, at the outside of the femoral artery. Just above this ligament several branches are given off which are distributed to the muscles and integuments of the pelvis and thigh. Three of these branches are called the *anterior, middle,* and *internal cutaneous* nerves, and are distributed to the integuments. The *saphenus* nerve is a branch of the anterior crural, that accompanies the femoral artery till the latter perforates the adductor tendon, after which it accompanies the saphena vein to the foot, giving off in its course branches to the integuments of the inner side of the leg, and to the upper and inner parts of the foot.

The *obturator* branch accompanies the obturator artery, and passing out at the obturator foramen, is distributed to the adductor and obturator muscles.

Sacral Nerves.

There are generally five pairs of sacral nerves, sometimes however there are six. Their anterior fasciculi unite with a part of the last dorsal, to form the *sacral* or *sciatic plexus,* which is situated at the side of the rectum, in front of the pyriformis muscle.

A number of branches are given off from the plexus, of which a few small ones go to the viscera and muscles within the pelvis. There are also given off the *two glutei* nerves which pass out at the sciatic notch, and are distributed to the glutei muscles; the *inferior long pudendal,* which winds around the tuberosity of the ischium, and is distributed to the integuments of the perineum; the *posterior cutaneous,* which is distributed to the integuments on the back of the thigh and leg; and the *superior long pudental* or *internal pudic,* which accompanies the internal pudic artery, and supplies the organs of generation and the perineum.

The *great sciatic*, or *ischiatic* nerve, also arises from the sacral plexus. It is much the largest nerve in the body, and passing out from the pelvis under the pyriformis muscle, it descends the back part of the thigh about half way to the knee, and then divides into two large branches, called *peroneal* and *popliteal*. In its course the sciatic gives off several branches to the integuments and muscles of the upper part of the thigh.

The *popliteal* nerve continues straight down to the middle and posterior part of the knee joint, and at the head of the tibia, divides into the *external peroneal* and the *anterior tibial*, the first of which descends upon the fibular side of the leg to the foot and ankle, and the last passes down in front of the interosseous ligament, and is distributed to the muscles upon the foot. The *posterior tibial* nerve is a continuation of the popliteal, and descending the back part of the leg along with the posterior tibial artery to the sole of the foot, divides into the *external* and *internal* plantar nerves which supply the sole. The *external saphenus* is also a branch of the popliteal, near the ankle it becomes superficial, and is distributed to the external ankle and foot.

The Sympathetic Nerve.

The sympathic nerve consists of two series or chains of ganglia placed on either side of the lateral part of the bodies of the vertebræ, and extending from the base of the cranium to the coccyx. With the exception of the neck, there is a ganglion for each intervertebral space; the ganglia are united to each other by nervous cords, and send off numerous filaments to the adjacent organs. Each ganglion is considered as a distinct nervous centre. This nerve supplies all the internal organs, and by means of plexuses is connected and distributed with all the other nerves of the body. Besides the ganglia situated along the spine, there are others placed about some of the large vessels of the abdomen, and also in other parts of the body.

In the cranium are five ganglia belonging to the sympathetic, viz. the *ganglion of Ribes*, situated upon the anterior communicating artery, and connected with the *carotid plexus* by means of a filament of the sixth nerve. The carotid plexus surrounds the carotid artery in the carotid canal, and is joined by the deep petrous branch of the vidian nerve;—the *lenticular ganglion*, situated between the optic nerve and the external rectus muscle, and communicating with the nasal nerve, a branch of the third pair, and with the carotid plexus;—the *spheno-palatine* or *Meckel's ganglion*, which gives off the vidian nerve, the deep petrous portion of which, as mentionned, joins the carotid plexus, and the superficial petrous joins the submaxillary ganglion; this ganglion also communicates with the superior maxillary nerve, and from it the palate, gums, and fauces, are supplied. A small oval ganglion, called the otic ganglion, is attached to the inner surface of the inferior maxillary nerve below the foramen ovale, from which communicating branches go to the superior and inferior maxillary, and to the vidian and tympanitic nerves, and which also supplies the tensor tympani, and the tensor palati muscles. A ganglion, called the ganglion of *Laumonier*, is also frequently found in the carotid plexus.

In the neck are three ganglia, called the *superior*, *middle*, and *inferior cervical*. The first is long and spindle shaped; it is situated along the sides of the third and fourth cervical vertebræ, behind the sheath of the vessels of the neck. Its branches are numerous, and communicate with the first, second, and third cervical nerves, with the carotid plexus and the second cervical ganglion, with the facial eighth and ninth pair, and the pharyngeal plexus, and with the cardiac ganglion, by means of the superior cardiac nerve.

The *middle cervical ganglion* is smaller and more flattened than the preceding; it is placed in front of the fifth or sixth cervical vertebra. In some instances it is wanting. It has, also, numerous connections; its branches join the anterior fasciculi of the third, fourth, and fifth cervical nerves, and also the superior and inferior cervical ganglia and the middle cardiac plexus.

The *inferior cervical ganglion* is placed near the head of the first rib and varies in form and size; it is mostly larger than the last. Like the two last, it also gives off numerous filaments; it is connected with the sixth, seventh, and eighth nerve, and from it proceeds the *inferior cardiac* nerve, which joins the middle cardiac nerve and the cardiac plexus.

The *cardiac plexus* is situated beneath the arch of the aorta. It is formed almost entirely by branches from the three cervical ganglia, the most of them coming from the middle one, or the middle cardiac nerve. It is a single plexus, formed by the nerves from both sides of the neck. Filaments from the *par vagum* and the *descendens noni* nerves are also blended with it. From this plexus the heart is supplied with nerves.

In the *thorax* are *twelve ganglia,* which are situated at the commencement of each intercostal space near the heads of the ribs. They are smaller than the cervical ganglia, and are connected with each other, and also with the anterior fasciculi of the spinal nerves.

The *great splanchnic nerve* is formed by filaments derived from the sixth to the tenth ganglion; it descends through the posterior mediastinum, and penetrating the diaphragm along with the aorta, forms on each side of the aorta a large ganglion composed of a number of smaller ones, called the *semilunar ganglions.*

The *small splanchnic nerve* is derived from the tenth and eleventh thoracic ganglia, after passing through the diaphragm, it joins the semilunar ganglion and the renal plexus.

The *solar plexus* is a network of nerves situated on the sides of the aorta, and extending downwards as far as the renal arteries. It is composed of the several filaments connecting the semilunar ganglion. From it proceed a number of smaller ganglia, which accompany the several arteries. The *diaphragmatic plexus* accompanies the phrenic arteries; the *superior coronary plexus* accompanies the corresponding artery to the stomach; the *splenic plexus* accompanies the splenic artery to the spleen, &c. The superior and inferior mesenteric, the renal plexuses, &c., are distributed in the same manner.

Along the sides of the lumbar vertebræ are placed the *lumbar ganglia*, generally five in number. They are also united to each other, and to the spinal nerves.

The *sacral ganglia*, mostly three in number, are situated on the anterior face of the sacrum. The lumbar and sacral ganglia unite with branches from the lumbar and aortic plexuses, and form the *hypogastric plexus*, from which all the pelvic viscera are supplied.

Physiology of the Nervous System.

The certebrum.—In the cerebral hemispheres, the highest and most important of the functions of the animal economy—those of the *mind*—are seated. In them the faculty of attention, or the power of directing the mind to impressions made on the senses, resides. The cineritious or gray matter which is found in the convolutions on the periphery of the cerebrum, is the portion which possesses these elevated functions. This gray matter, wherever found, whether in the brain and spinal marrow, or in the ganglions along the course of the nerves, is con sidered as a nervous centre, or the generator of nervous influence, while the white or medullary matter carries this influence to the different parts of the body.

The more numerous and complex the convolutions in general, the greater the degree of intelligence.

In infants they are imperfectly developed, and their increase is proportionate to the mental improvement; if their growth be arrested by any cause, the mental powers are feeble. Idiots, besides having small brains, have but a limited development of the convolutions. The arrangement of the brain into convolutions admits of a large surface of cineritious matter in a small space. It also allows of a more ready access to the blood-vessels on the one side, and a more free communication on the other, with the fibres by which its influence is distributed. The entire surface of a human cerebrum of average size, when the convolutions are unfolded, has been estimated to be equal to about 670 square inches.

The hemispheres possess but little or no sensibility, they may be wounded, and partially or entirely removed, without giving rise to pain; when severely injured, however, a state of stupor, attended by general functional derangement, mostly results. Instances have occurred in which portions of the cerebrum have been removed without destroying life or impairing the intellect.

The cerebellum.—With respect to the functions of this portion of the brain there is much diversity of opinion; the majority of observers, however, agree in regarding it as the seat of the animal or lower propensities. It is urged in support of this view, that in individuals who have given free indulgence to their passions, the relative size of this organ is much increased.

On the other hand, some contend that is has but little or nothing to do with these propensities, but that its function is to regulate and harmonize the muscular movements, especially those of a voluntary character. Like the cerebrum, the cerebellum is void of sensibility.

The *medulla oblongata.*—Placed intermediate between the brain and spinal marrow, the medulla oblongata serves as a medium through which they act on each other. The corpora pyramidalia, or anterior pyramids, connect the motor fibres of the cerebrum with the anterior lateral columns of the spinal cord; near the lower extremity of the medulla oblongata, these fibres decussate, or cross from side to side, a large portion of those, coming from the right side of the cerebrum, passing over to the left side of the cord, and *vice versa.* Hence the frequent occurrence of paralytic affections on the side of the body opposite to that affected in the brain. It is also alleged that, besides being the point at which sensation terminates and excitement to motion begins, the medulla oblongata possesses the power of originating motion in itself, independent of the cerebrum, and that it presides, especially, over respiration and deglutition.

The *medulla spinalis.*—The functions of the spinal cord are to convey nervous influence to and from the brain, and also to originate nervous influence independently of the brain. All the nerves of the body are united into one common trunk in the white or medullary matter of the cord; though each filament of every nerve runs a separate and distinct course from its starting point to its termination. Hence the brain influences not only the nerves at the base of the cranium, but all the spinal nerves through the medium of the spinal cord. As previously mentioned, the spinal nerves arise by two roots from the anterior and posterior columns of the spinal cord; a part of the fibres of the posterior root, which is distinguished by having a ganglion on it, pass on to the brain conveying impressions to this organ; and part terminate in the gray matter of the spinal cord, conveying impressions to it. This root is called the *sensory* root; it is also the *afferent* root. The anterior is the *efferent* or *motor* root; part of its fibres come from the brain, conveying *voluntary* motion, and part of them originate in the gray matter of the spinal marrow. To these fibres of both roots, which appear to act independent of the brain, constituting with the gray matter of the spinal cord a distinct nervous centre, the term *reflex system* has been applied.

Thus it will be seen that each nerve has four sets of fibres: one set, called *sensory*, passing upwards to the brain, conveying sensations to that organ; a second set, called motor, conveying the influence of volition and emotion *from* the brain; a third set, called *excitor*, terminating in the spinal cord and conveying impressions to it; and a fourth set also of *motor* fibres, conveying the motor influence from the spinal marrow to the muscles.

In the *reflex system*, or that of which the spinal cord is the centre, it will be understood that the cord has the power of reflecting the action of the sensitive upon the motor nerves, without itself possessing *sensation*, that being a faculty which belongs exclusively to the brain.

The *sympathetic nerve.*—This nerve is both motor and sensory; it exercises a controlling influence over the involuntary functions, and being connected also with the cerebro-spinal system, it brings the organic functions in relation to the animal.

Sensation.

Sensation is defined to be "the perception of an impression;" it is to the brain alone that this faculty belongs, hence the term *sensorium* is often applied to this organ. There are two kinds of sensations, one internal, the other external; the first arise from impressions made within the body, as the sensations of hunger or thirst, or such as arise from some temporary want of the system; the second are those which arise from impressions made on the external surface of the body, as the sense of touch or sight. Sensation occurs in the brain, and not in the part impressed.

Sensations are likewise divided into *general* and *special*. By general sensation, which is distributed all over the body, we feel those impressions made by surrounding objects, which produce the various modifications of pain and pleasure, variations of temperature, and the sense of contact and resistance. By special sensation is understood that which arises from impressions of a peculiar character upon nerves which are adapted to receive them alone. Each nerve of special sensation requires its own peculiar stimulus to call it into action, and is consequently incapable of taking part in the action of another. Thus light is required for the eye, sound for the ear, &c. Nerves of *special* sensation have no general sensibility, they may be wounded without causing the individual any pain.

There are five special senses, viz. touch, taste, smell, hearing, and seeing.

Sense of Touch.

This sense enables us to become acquainted with the hardness or softness, the roughness or smoothness, shape, size, and weight of a body. The idea of *resistance* would seem to be the only idea conveyed to the mind by the sense of touch, as by it these properties of bodies are made known. The sense of touch is more highly developed in the lips, tip of the tongue, and the palmar surface of the extremities of the fingers, these parts being abundantly supplied with nerves of general sensation, than in other parts of the body; it is also more generally distributed throughout the animal kingdom than any of the other senses. The nerves of touch are the posterior roots of the spinal nerves, and portions of the fifth and eighth pairs of cranial nerves; these are also the nerves of *general sensation;* they are distributed to the papillæ of the skin, and covered by the epidermis to protect them from too violent external impressions from external bodies. All bodies to be cognizable to the sense of touch, must be brought into contact with the sensory surface; the only exception to this rule is with respect to the sense of temperature, for which, in the opinion of some physiologists, there is a special set of nerves.

Sense of Taste.

The mucous membrane of the tongue and fauces is the organ of taste, the anatomical character of which has been previously described.

As in the sense of touch, so in that of taste, the substances to be examined must be brought in contact with the organ. When substances having a strong savour are brought in contact with the tongue, the papillæ become erect and turgid, giving to the surface of the organ a decided roughness. For the exercise of this function it is necessary that the substance to be tasted should be soluble; otherwise the feeling of contact merely is excited. Impressions of taste remain longer than those of the other senses; though the after-taste may be different from the original.

There is no special nerve of taste; the tongue being supplied by the fifth and eighth pairs, which would seem to convey impressions of taste as well as give to the tongue its general sensibility. The first of these nerves is distributed more to the front, and the second more to the back of the organ; those impressions which produce nausea are conveyed by the latter. The motions of the tongue are performed through the ninth pair, though this nerve has nothing to do with the sense of taste; as its division causes loss of motion of the organ, without at all impairing the special sense.

Both the senses of touch and taste are deadened by cold air; the latter is also considerably impaired by any injury of the sense of smell.

Sense of Smell, or Olfaction.

The *nose* is the *organ* of smell; it consists of two portions, one external, projecting upon the face, the other an internal cavity. The external portion is formed by the nasal bones, the nasal process of the superior maxillary bones, by five cartilages, and by the integument. The bones have been already described. Two of the cartilages are placed on either side, and one in the middle; the latter constitutes the cartilaginous septum between the nostrils, and is thick, flat, and triangular. The lateral cartilages are also triangular, and articulate above and behind with the bone, in front with the septum, and below with the alar cartilages. The alar cartilages form the lower part of the nose, called the nostrils; they are irregularly semi-eliptical, and keep the nostrils open. The mucous membrane lining the nose is thick, soft, and red; it is termed pituitary or Schneiderian, and is continuous with the mucous membrane of the mouth, Eustachian tube, lachrymal canal, and frontal sinus. The hairs situated at the entrance of the nose are called *vibrissæ.*

By the sense of smell we are made acquainted with the odorous particles of bodies suspended or dissolved in the atmosphere. Being seated in the nose, at the entrance of the respiratory passages, it serves as a protection against the introduction of injurious matters. Another use of this sense, and the principal one, is to aid the impression of taste in conveying intelligence of the properties of food. The sense of smell is limited to that portion of the mucous or Schneiderian membrane of the nostrils which covers the *superior* and part of the middle turbinated bones, the *olfactory* nerve being distributed only to this portion. Hence this region is called the olfactory region. The advantage of having the sense situated high up in the nostril, is to protect it from mechanical injury, and also from the contact of too cold or too dry air, both of which impair it very materially. Odorous particles are brought in contact with the Schneiderian membrane by the act of inspiration, and are so minute that they can not be detected by the most delicate experiments. The fifth pair of nerves give to the mucous membrane of the nose its general sensibility.

The sense of smell, like that of taste, is not an intellectual one; it is however, susceptible of cultivation; individuals by it are frequently capable of recognizing others. It is usually much more acute in the lower orders of animals than in man.

Sense of Hearing, or Audition.

The *ear* is the organ of *hearing.* It consists of three parts; the *external ear*, the *middle ear* or *tympanum*, and the *internal ear.* The external is composed of the *pinna,* which is the movable part on the side of the head, and the meatus, or canal, a passage leading from the pinna. The outer rim of the pinna is called the *helix,* within which is a prominence called the *anthelix.* At the upper part this prominence divides, leaving a space termed the scaphoid fossa. The deep, central cup within the anthelix, is the *concha.* A small eminence, situated at the end of the helix, and in front of the concha, and containing a small tuft of hair, is the tragus—so named from its resemblance to a goat's beard. Opposite to this eminence, and below the concha, is the *antitragus.* The pendulous portion is the lobe; it consists of cellular and adipose tissue. The oval, elastic plate of fibro-cartilage is attached in front, by the anterior ligament to the zygomatic process, and behind, by the posterior ligament to the mastoid process. Several small muscles enter into the composition of the pinna, but as they are, comparatively, of but little importance, their description may be dispensed with.

The *meatus* is about an inch in length; it is a bony canal lined by cartilage, narrow in the middle, and curved downwards. The skin lining it is studded with hairs, and glands which secrete wax or cerumen.

The *middle ear*, or tympanum, is an irregular cavity situated in the petrous portion of the temporal bone, and filled with air which enters by the Eustachian tube; it contains a chain of small bones, and openings into the mastoid cells. In front it is bounded by the *membrana tympani*, or drum of the ear.

The *membrana tympani*, or drum, is a thin, transparent, oval membrane which separates the external ear from the cavity of the tympanum. It is placed obliquely across this cavity, and is slightly convex internally, and concave externally. It consists of three lamina; the external is continuous with the cuticle, the internal with the mucous membrane lining the cavity, while the middle is strong and fibrous, and attached by its circumference to the bone.

The middle ear has two orifices communicating with it; one, the *fenestra ovalis*, leading to the vestibule; the other, the *fenestra rotunda*, opening into the cochlea; both of these, however, are closed by membranes to prevent the escape of fluid contained in the inner chambers. The Eustachian tube is a straight canal, about two inches in length, which empties into the pharynx; its commencement is bony, and its termination cartilaginous.

The bones of the tympanum are four in number, viz. the *malleus*, or hammer, which is imbedded in the tympanum; the *incus*, or anvil, somewhat resembling a molar tooth; the *orbiculare;* and the *stapes* or stirrup. These bones articulate with each other in the order named, and are also connected by three small muscles, called the *tensor* and *laxators tympani*, and the stapedius. The contractions and relaxations of these muscles move the bones, and relax or make tense the tympanum.

The *internal ear*, or *labyrinth*, is composed of three parts, viz. the *vestibule*, the *semicircular canals*, and the *cochlea*.

The *vestibule* is a small triangular cavity situated within the wall of the tympanum; into it behind, the semicircular canals open by five orifices, and in front the cochlea by a single one. The fenestra ovalis is on its outer wall, and on its inner are several small holes for the passage of a portion of the auditory nerve. The aqueduct of the vestibule also opens into it.

The *semicircular canals* are three bony passages, the extremities of which open into the vestibule by five orifices. Two of these canals are vertical, and one horizontal.

The *cochlea* resembles a snail shell, and forms the anterior portion of the labyrinth; it consists of a bony and gradually tapering canal about an inch and a half in length, which makes two turns and a half spirally around an axis called the *modiolus.* The canal of the cochlea is divided into two passages by a bony and membranous plate, called the lamina spiralis; these passages communicate with each other at the apex of the cochlea. The *aqueduct* of the cochlea opens by one extremity into the upper part of the canal, and by the other upon the inferior surface of the petrous bone.

The *membranous labyrinth* has the same form as the bony vestibule, cochlea, and semicircular canals, which cavities it lines; it consists of several layers, and contains a limpid fluid called after Cotonius.

The auditory nerve divides in the internal meatus into two branches, which ramify through the membranous labyrinth and terminate on the inner surface of the membrane.

The *sense of hearing* is that function by which the mind takes cognizance of the vibratory motions of bodies which give rise to the phenomena of sound. These vibrations may be communicated to the ear through the air, or through the intervention of some solid substance brought into contact with the organ of hearing.

The precise function of all the parts of the ear is not known.

The function of the *external ear* is to collect and concentrate sonorous vibrations or sounds, and conduct them inwards.

The use of the *membrana tympani* is to receive the sounds and transmit them to the chain of bones, and also to modify their intensity. It is not, however, essential to hearing as it may be perforated or destroyed without very materially impairing the sense.

The *chain of bones* serves to conduct the sounds across the cavity of the tympanum to the internal ear.

The *tympanum* isolates the chain of bones, and allows free vibration to the membranes at each end of it. The use of the *Eustachian tube* is to admit air into the cavity of the tympanum, which prevents undue tension of the membrane, by rendering the pressure on both sides equal; and to carry off the secretions of the middle ear which it discharges into the throat.

In regard to the functions of the *internal ear*, or *labyrinth*, but little is known; in it the sounds reach the auditory nerve, and are thus transmitted to the brain.

The idea of the *distance* and *direction* of sounds is mostly acquired by habit. The *acuteness* of hearing varies very much in different individuals, and may be much increased by cultivation.

Sense of Vision.

By this sense we are enabled to perceive the form, size, color, position, &c., of the bodies which surround us. The medium through the agency of which this is accomplished, is called *light*. The eye is the organ of sight.

The *globe*, or ball of the eye, is a spherical body about an inch in diameter, from before backwards and somewhat less transversely. It is surrounded by a fibrous membrane continuous with the sheath of the optic nerve behind.

The *sclerotic coat* is a dense, white, fibrous membrane, which invests about four-fifths of the eye, giving to it its form, and serving for the attachment of the muscles. Behind it is perforated by the optic nerve. In front it receives the cornea by a circular, bevelled edge.

The *cornea* is the transparent, projecting portion which fills up the circular opening in the anterior part of the sclerotica; it constitutes about one-fifth of the eye-ball, and is composed of several lamina, or layers.

The *choroid coat* is a thin vascular membrane which lines the sclerotica, and is of precisely the same extent as that tunic.

It consists of three layers, and is filled with black coloring matter called pigmentum nigrum.

The *ciliary processes* are situated at the anterior portion of the choroid coat surrounding the lens. They consist of seventy or eighty short triangular folds.

The *iris* is a thin circular membrane, varying in color in different persons, hence its name, forming a vertical septum, or partition, between the anterior and posterior chambers of the eye. In its centre is an opening, called the pupil. Its posterior surface is in contact with the ciliary processes, and thickly coated with pigmentum nigrum. Its external border is attached to the ciliary ligament at the base of the ciliary processes.

The *retina* is a soft white membrane, situated within the choroid coat; it extends from the optic nerve behind to terminate in an irregular edge around the centre of the cilia in front. Internally it is in contact with the vitreous humor. It is a nervous structure, and generally considered as an expansion of the optic nerve. This nerve enters the ball of the eye on the inner side of its axis; and through it the central artery of the retina enters, and after passing through the vitreous humor, is distributed to the retina and lens. The point where the optic nerve is connected with the retina is incapable of vision. Immediately at this point, also, there exists upon the retina a yellow spot, called after *Sœmmering*.

A thin membrane between the choroid coat and the retina, is the *membrana Jacobi*.

The *aqueous humor* is a transparent, albuminous fluid, filling the chambers of the eye. The anterior chamber is the space between the cornea in front, and the iris and pupil behind; the posterior chamber is between the iris and pupil in front, and the crystalline lens and ciliary processes behind. Both chambers are lined by a delicate membrane which secretes the aqueous humor. If the aqueous humor is by accident discharged, it will be replaced again without loss of sight.

The *crystaline lens* is placed immediately behind the pupil, and is surrounded by the ciliary processes; it is a double convex lens, with the posterior surface more convex than the anterior. It is soft and transparent, and composed of concentric laminæ or layers which increase in hardness as they approach the centre. The lens is invested by a transparent membrane, called its capsule. In childhood it is spherical, and in old age it is flattened.

The *vitreous humor* is more dense than the aqueous, being of a jelly-like consistence. It is transparent and globular in form, constituting the principal bulk of the eye. It is held in cells formed by a delicate membrane, called the hyaloid. If the vitreous humor is let out, it cannot be restored, and the eye is entirely destroyed.

Fig. 30.

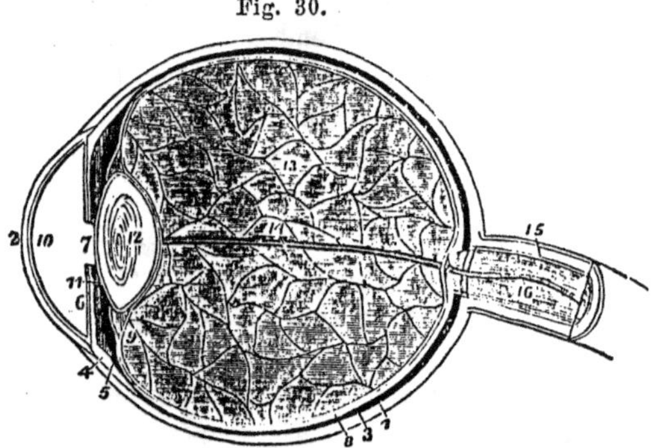

Fig. 30. An antero-posterior vertical section of the eye. 1 The sclerotic coat; 2 the cornea; 3 the choroid coat; 4, 5 the ciliary body and processes; 6 the iris; 7 the pupil; 8 the retina; 10 anterior chamber; 11 posterior chamber; 12 crystalline lens; 13, 14 vitreous humor; 15 central artery of the retina; 16 optic nerve.

The *eyebrows* are elevations of integument, covered with stiff short hairs: they protect the eye from too vivid light and from the perspiration that accumulates on the forehead.

The *eyelids* or *palpebræ*, are composed of skin, muscular fibres, and cartilage; the latter, called *tarsal cartilages*, are small and crescentic in shape, and placed in the edges of the lids; they serve to preserve the shape of the lid. When the lids are in contact, they have a triangular canal between them.

The *cilia* or *eyelashes* are short, curved hairs placed in the margins of the lids to protect the eyeball. A mucous membrane, called the *conjunctiva*, lines the lids, and also extends over the anterior surface of the ball.

The *Meibomian glands* are small bodies, twenty or thirty in number, placed between the conjunctiva and the inner surface of the lids. They discharge a viscid fluid by numerous orifice along the edges of the lids, and their secretion prevents the overflow of tears at night.

A small red elevation in the internal angle of the eye, about the size of a grain of barley, is the *caruncula lachrymalis.* It consists of a number of small glands.

The *lachrymal gland* is situated at the upper and outer portion of the orbit; it is of a light pink color, and about the size of a filbert. It secretes the tears, which are discharged through ten or twelve ducts opening along the edge of the upper lid. By this secretion the eye is constantly kept moist.

The *lachrymal canals*, one for each lid, commence by a very minute orifice, called *puncta lachrymalia*, near the inner angle of the eye. They conduct the tears into the lachrymal sac. The *lachrymal sac* is the enlarged upper extremity of the *nasal duct.* This latter is a canal, about three-fourths of an inch in length, passing downwards and backwards from the inner angle of the eye through the bones of the face to the inferior meatus of the nose. It carries off the tears.

The movements of the eye are performed by six muscles, which arise from the walls of the orbit, and are inserted into the sclerotic coat. Four of them are straight, and are called the *superior*, *inferior*, *internal*, and *external:* and two are *oblique* —a superior and inferior. The *superior oblique* muscle plays over a tendinous pully attached to the upper margin of the orbit.

In order to comprehend fully the functions of the different parts of the eye, a knowledge of the laws of light and optics is necessary, for which the reader is referred to the works of natural philosophy, as these subjects belong rather to natural philosophy than to physiology.

The use of the sclerotic coat is to give form to the body of the eye, and also to protect the inner and more delicate parts.

The *choroid coat* serves chiefly to transmit the nerves and blood-vessels.

The *pigmentum nigrum* on the inner surface of the choroid coat absorbs the rays of light after they have made their impressions on the retina.

The use of the iris is to regulate the quantity of light admitted through the pupil, which it is enabled to do by its power of contraction and expansion.

The *aqueous humor*, *crystalline lens*, and *vitreous* humor, are transparent media through which the rays of light pass to reach the retina; their office is to refract rays of light so that they will fall upon the retina in the most favorable manner; the most of the refracting power is possessed by the lens.

The use of the retina is to receive the impressions of light, and transmit them to the brain.

In near-sighted persons the refracting power of the eye is too great, the rays of light being brought to a focus before reaching the retina. This defect is remedied by the use of a double concave lens. In far-sighted persons, on the contrary, there is not sufficient refractive power, and the focus is formed behind the retina. To obviate this defect, convex lenses should be used to concentrate the rays. The former defect, that of near-sightedness or *myopia* commonly occurs in young persons, and is often corrected by age; the latter, far-sightedness or *presbyopia*, is most frequently met with in persons somewhat advanced in years.

The fifth pair of nerves furnishes the eye with general sensibility; the optic nerve being a nerve of special sensibility only.

INDEX.

R.

S.

T.

UNIVERSITY OF MICHIGAN
3 9015 00660 7629

www.ingramcontent.com/pod-product-compliance
Lightning Source LLC
LaVergne TN
LVHW011221110826
845150LV00006B/1493

* 9 7 8 1 4 2 5 5 1 6 1 8 5 *